THE LEARNING

OF

MATHEMATICS

ITS THEORY AND PRACTICE

THE LEARNING

OF

MATHEMATICS

ITS THEORY AND PRACTICE

Twenty-First Yearbook

THE NATIONAL COUNCIL OF TEACHERS
OF MATHEMATICS

WASHINGTON, D. C.   1953

COPYRIGHT © 1953 BY
THE NATIONAL COUNCIL OF TEACHERS
OF MATHEMATICS, INC.

Second Printing 1955
Third Printing 1961
Fourth Printing 1964
Fifth Printing 1967

Correspondence relating to and orders for
additional copies of the Twenty-first
Yearbook and the earlier yearbooks should
be addressed to

THE NATIONAL COUNCIL OF TEACHERS
OF MATHEMATICS
1201 SIXTEENTH STREET, N. W., WASHINGTON, D. C. 20036

Manufactured in the United States of America

Dedicated to
William David Reeve
in recognition and appreciation of
his great service to
The National Council of Teachers of Mathematics
as Editor
of the Third through the Twentieth Yearbooks

Preface

How does the human brain and nervous system acquire its store of mathematical knowledge? How does the human organism use this store of knowledge once it has acquired it? These are fundamental questions to which the answers can be of great aid in the improvement of instruction in mathematics. Although comparatively little is known about the answers, the little that is known should be studied by every teacher of mathematics on every level of instruction. It is the purpose of this Yearbook to provide some of this information, and indicate sources for further study.

The title of the Yearbook tells the organizational pattern. Each chapter discusses an important aspect of learning, giving the most modern theory and research, and then applies this theory to concrete learning situations. In this way it is hoped that the classroom teacher will not only be given concrete suggestions, but also a theoretical background upon which to create his own provisions for better learning of mathematics. Insofar as teachers find the materials presented here of value in providing better classroom learning situations, the book will have succeeded in its purpose.

This Yearbook was inaugurated by the first Yearbook Planning Committee of the National Council of Teachers of Mathematics. The members were Mr. Walter H. Carnahan, chairman, Miss Veryl Schult, and Mr. F. Lynwood Wren. The editor is greatly indebted to them for their help and encouragement in getting the book organized and under way.

The arrangements for editing final copy, printing, securing permission to use materials, and other business details were numerous and complex. The efficient aid of the executive secretary, Mr. Myrl H. Ahrendt, in all these matters is gratefully acknowledged.

The authors received much help and guidance from their colleagues in their various institutions of learning. This help, especially from psychological departments, is deeply appreciated and has contributed much to make the book authoritative. For permission to reproduce figures and printed material acknowledgment is hereby made to the following persons and companies: E. Heidbreder; Elizabeth M. Thorndike; Aaron Bakst; H. G. and Lillian R. Lieber; University of Chicago; National Society for the Study of Education; Princeton University Press; Columbia University Press; Harcourt, Brace and Company; New York Academy of Sciences; American Psychological Association; the Macmillan Company; The Science Press: Appleton-Century-Crofts; Harvard University Press; Houghton Mifflin Company; Duke University Press; Henry Holt and Company; John Wiley and Sons; The Journal Press; Journal of Educational Research; Longmans, Green and Company; Thomas Y. Crowell Company; World Book Company; Harper and Brothers; University of Chicago Press; Institute of General Semantics; The Alfred Korzybski Estate; and Oxford University Press.

HOWARD F. FEHR
Editor

Contents

PREFACE .... vii–viii
I. THEORIES OF LEARNING RELATED TO THE FIELD OF MATHEMATICS 1
    Howard F. Fehr, Teachers College, Columbia University, New York City
II. MOTIVATION FOR EDUCATION IN MATHEMATICS 42
    Maurice L. Hartung, University of Chicago, Chicago, Illinois
III. THE FORMATION OF CONCEPTS 69
    Henry Van Engen, State Teachers College, Cedar Falls, Iowa
IV. SENSORY LEARNING APPLIED TO MATHEMATICS 99
    Henry W. Syer, Boston University, Boston, Massachusetts
V. LANGUAGE IN MATHEMATICS 156
    Irvin H. Brune, State Teachers College, Cedar Falls, Iowa
VI. DRILL—PRACTICE—RECURRING EXPERIENCE 192
    Ben A. Sueltz, State Teachers College, Cortland, New York
VII. TRANSFER OF TRAINING 205
    Myron F. Rosskopf, Teachers College, Columbia University, New York City
VIII. PROBLEM-SOLVING IN MATHEMATICS 228
    Kenneth B. Henderson and Robert E. Pingry, University of Illinois, Urbana, Illinois
IX. PROVISIONS FOR INDIVIDUAL DIFFERENCES 271
    Rolland R. Smith, Public Schools, Springfield, Massachusetts
X. PLANNED INSTRUCTION 303
    Irving Allen Dodes, Stuyvesant High School, New York City
XI. LEARNING THEORY AND THE IMPROVEMENT OF INSTRUCTION—A BALANCED PROGRAM 335
    John R. Clark and Howard F. Fehr, Teachers College, Columbia University, New York City

1. Theories of Learning Related to the Field of Mathematics

Howard F. Fehr

WAYS OF STUDYING LEARNING

There are a number of ways to study the learning process of human organism. One is purely physiological, that is to study learning as physical reactions of the brain, the nervous system, the glands, and the muscles, as they are acted upon by physical stimuli. Another method, partly physiological and partly observational, is to study the way the organism reacts in various situations so as to abstract common elements called laws of learning. It is recognized that physical changes are taking place in the organism, and that some of these physical changes can be ascribed to certain actions and reactions of the organism in particular situations. But the general explanation of the reaction of the organism is given in terms of the situation, and not in terms of physical changes within the organism. This procedure is followed by psychologists. A third method is to ignore all internal physical changes and to describe learning purely in terms of introspection and logical considerations. All three methods have provided and are providing new insight into human behavior, but recently the psychological investigations have given the most promise of help to the teacher.

What do we know about the physical behavior of the brain? In the first place, it is composed of more than 10 billion nerves which are connected by an exceedingly complex network. These neurons consist of a center or cell body, from which run fibers of two types, axons which are single strands of various lengths, and dendrites, which are ramified short fibers. Impulses travel along these nerve fibres at rates which have been measured to vary from three feet to 300 feet per second. The impulse is relayed from one neuron to another by a synapse, and the flow of the impulse is in one direction only. The response of each fiber is an "all or none," that is, if it is not sufficiently agitated there is no response, but at a certain degree of stimulation the whole response goes forward. The amount of stimulation necessary for the "all or none" response varies also from neuron to neuron. Hence any overt action of an individual, and all actions, involves many, many nerve fibers, and is dependent upon the number of stimulated neurons.

We also know that certain areas of the brain are related to certain functions such as sight, information storage, control of the sympathetic nervous system, and emotional behavior, and that damage to these parts of the brain interferes with the corresponding functions. There is also evidence that in time, certain parts of the brain can take over the functions of other damaged parts.

But how the physical behavior of this vast network of nerves in the brain and nervous system produces the response $a^2 - b^2$ is $(a - b)(a + b)$ is totally unknown. How the cells get their information, how they transfer it from the sign of $a^2 - b^2$ to one area of the brain, to another area, to an ultimate response from the organism of $(a - b)(a + b)$ is a deep, dark secret. Further, any attempt to study the physical behavior meets with many obstacles. To open the brain to observation is usually accomplished by destroying the very nerves we would study. Further, the nerve cells, axons, and dendrites are exceedingly minute objects, and to see a synapse at the end of a nerve fiber is exceedingly difficult. At present, explanation of human actions in terms of physical phenomena within the brain seems very, very remote (23).[1]

Hence psychologists have resorted to experimental and observational procedures to explain what the human brain is, and what it does. They create certain situations and observe under as controlled conditions as possible the behavior of the organism, and describe the operation of human learning by the various behaviors that take place. Thus human learning is defined as a change in behavior acquired through an experience. The learning is usually directed toward specific goals through organized patterns of experience. In order to clarify our concept of change in behavior, we give several examples from the mathematical field.

EXAMPLES OF CHANGE IN BEHAVIOR

When a student enters a beginning algebra class and is asked, "What are the two numbers of which the sum is 6, and the differ-

[1] In this book the symbol $(x:y)$ will be used to refer to page y of reference x in the numbered list at the end of each chapter.

ence is 1?" he *behaves* as follows: Try 1 and 5; the difference is 4, no. Try 2 and 4; the difference is 2, no. Try 3 and 3, no. Maybe there is no answer; try again; 2 and 4; 3 and 3. Oh! Maybe I can use fractions; 1½ and 4½, no; 2½ and 3½; there it is. He has solved the problem, he has reasoned, and he has exhibited a type of behavior in a given situation, but it is not the goal behavior you will ultimately expect from his instruction in algebra. Now let us repeat the same problem three months later. If the student reacts in the same manner as above, he has not learned anything new in this situation. If, however, he behaves as follows: Two numbers x and y; sum, $x + y = 6$; difference $x - y = 1$; add $2x = 7$, $x = 3½$ and $y = 2½$, then his behavior has decidedly changed; he has learned a new mode of action. His mind proceeds in a manner entirely different from before. We should set up our goals of learning in the mathematical field, in terms of all desired changes in behavior with reference to numerical, spatial, quantitative, and logical situations.

Another example. At the start of the year in plane geometry, you give the following hypothesis: A triangle has sides 2 in., 3 in., and 4 in. The middle points of sides 2 in. and 3 in. are joined by a straight line segment. Then you ask, "How long is this segment?" The student responds by using his ruler, a pair of compasses, and paper, actually constructing the triangle and segment and measuring the latter. Assuming careful work, the student responds, "I measure it to be 2 inches." His behavior in this case is a result of his past experience. Three months later you confront the student with the same problem. If he has learned, his response now is solely the result of an inner brain reaction. He says, "It is 2 inches, since it must be one-half the length of the third side." Thus he has had a complete change in behavior from one involving perceptual-motor skills to one involving purely concept-relationship.

One task in education is to create such experiences and situations that will enable a student to reconstruct his behavior towards goals desired by both himself and the teacher. When we have accomplished this, we shall have improved our instruction.

Learning thus becomes a developmental process. It is change in behavior brought about through brain action or thinking. It comes about through facing situations that call for making discoveries, abstractions, generalizations, and organizations in mathematics. It is problem-solving, for without a problem felt by the organism, and motivation toward the solution of the problem, there will be little learning of mathematics. On this most psychologies of learning agree. The disagreement arises in the theoretical explanation of how the solution comes about.

THE NATURE OF INTELLIGENCE

To what extent is it possible for human beings to change their behavior? It is quite common in academic circles to hear such expressions as, "He does not possess enough intelligence to learn mathematics," or "He is a highly intelligent individual." Intelligence as used in these expressions is that quality which permits an individual to adapt himself successfully to a given situation. This was one of the earliest definitions of intelligence. However, if a dog adapts himself to a household in a manner to get good care, we do not say the dog is intelligent (in the sense we apply the word to human behavior). There is more to intelligent action than mere adaptation.

Binet in his early work on testing used the ability to make judgments as the best description of intelligence. To this end he constructed many tests devised to measure the ability to make judgments or choices. This has culminated in the construction of many types of mental tests, and we could describe intelligence as that faculty, or quality, or characteristic which is measured by the intelligence tests. Intelligence would then be the ability to perform mental tasks, to remember, to make generalizations, to form relationships between concepts, and to deal with abstract ideas. The amount of intelligence would be measured by the degree of difficulty of tasks completed, of their complexity, of their abstractness, and of the speed and lack of interference with which the tasks are completed.

Dewey in all of his writings has concerned himself with the nature of human intelligence. A brief summary of his concept would be: Intelligence is acting with an aim; it is purposeful activity. The activity must at all times be controlled by a perception of all the facts in a given situation and their relationship to each other. Even more important in intelligence is the capacity to refer present conditions to future desired goals and conversely to refer the goal to the present conditions. Thus to refer counting to acquiring of addition facts (as an elimination of a needless time-consuming process) and to refer the facts back to counting is intelligence. If we act without knowing the consequences of our acts or even considering them, we are unintelligent. A shot in the dark is not intelligent action. (This is not to be confused with acting on a hunch; a hunch is usually related to the goal.) If we make a guess at an answer as a loose stab, and not as a related action, we may be exhibiting some intelligence (goal-directed action) but it is very imperfect. If, however, we act with an aim toward changing our behavior to a new desired pattern which is perceived as desirable, we are making intelligent action. "Intelligence is the power to understand things in terms of the use made of them" (3).

It is in this sense that intelligence is the ability to solve problems —to think—to learn. And this is more than merely an ability to think in terms of abstractions which is one kind of intelligence. It is also the ability to grasp relations in physical or concrete setups (situations) and to see how to readapt these for more useful purposes. This has been referred to as a practical or mechanical aspect of intelligence. A technologist has a different type of intelligence than a theoretical scientist. He foresees future conditions in terms of concrete situations rather than abstract relationships. His type of intelligence is very important in modern society and should be developed. It may be characterized in one way by a space-perception activity as contrasted with a deductive propositional activity. Another type of intelligence recognized by Thorndike (16) is social intelligence. This is the power to understand people, to get along with them and to lead them. It involves personality traits and actions between humans which relates present conditions to future desired states of happy, cooperative living, and vice versa.

In the learning of mathematics, the power with which an individual can make generalizations, abstractions, logical organizations, and relate these to purposeful action, determines his ability to progress. As teachers of mathematics, we are interested in this phase of intelligence. However, as teachers of children, of young men and women, we are certainly interested in the mechanical and social type of intelligence, and hence must consider all these types in our study of learning.

A LEARNING SITUATION

To study how we learn, consider your solution to the following problem: A man in a department store noticing the escalator in motion, raises the question, "How many steps are there in the escalator between the floors?" He walked down the escalator as it was in motion, timing the distance between floors. When he reached the lower floor having walked down 26 steps, it took him 30 seconds; similarly, when he walked down 34 steps, it took him 18 seconds. What is the answer to his question?

What answer did the man find? How did he find it? If you, the reader, are interested in these questions, if you really want to find the answers, you are in a learning situation. All your past experience in mathematics has created a mental set and the type of problem gives sufficient motivation to send you into action toward the solution. You now use your previous learning to find a solution. —(Before reading further, stop and seek your solution, keeping a diary of every move you make. Then you can study your method of obtaining a solution or how you learned in this situation.)

You may have gone directly to the solution of the problem on your first trial by applying a technique previously learned. In this case you did not learn, you did not need to learn, you merely recalled a previous learning. But you may have proceeded in one of the following manners: The difference between the numbers of steps walked during each of the two trials was 8; the difference in times was 12 seconds; but what relation has this to the problem? Here many a student would cease learning because he would have reached a block in his reasoning without sufficient drive to go on. However, another student would say, "Is there any relation between the difference in times and difference in steps traveled? Oh yes, there is a relation to the motion of the escalator and the steps move at a rate 8 steps per 12 seconds or $\tfrac{2}{3}$ steps per second. Now does this help me?" Again the student may be blocked or he may return to the problem and think, "In 30 seconds then, the escalator will move 30 × $\tfrac{2}{3}$ or 20 steps and the man move 26

steps. Aha! There are 46 steps in the escalator between floors."
Now many a student would stop having secured satisfaction with
the answer. But, if he is to have better learning he will check and
analyze his thinking as follows: "Let me see if this is so. If there
are 46 steps and I go down 34 of them, the escalator must move
12 steps. At a rate of ⅔ steps per second, it takes 12 ÷ ⅔ or 18
seconds and that's right." At this point, having checked a hypothesis, the student may again cease his learning. He has all that he
desires. But a still better learner will say: "Now let me see how I
solved the problem. First I found the rate at which the escalator
moved, then I found the number of steps the escalator moved in
30 seconds, and I added to this the number of steps walked. This
is the number of steps in the escalator. In situations where I
know two distances, and two corresponding times, I had better
try first to determine a rate."

This whole learning leads to a change in behavior. When confronted with a similar situation the student will now act differently
from what he did in solving this problem. Of course, there are
several ways of solving the problem besides this method of arithmetic. This method was shown to illustrate how we learn.

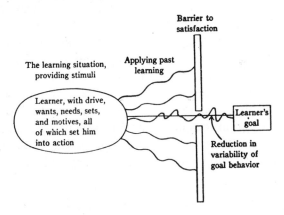

The accompanying diagram similar to that given by Dashiell (2)
can be taken as the starting point in the study of how we learn.

At the start of learning or readjustment of behavior, there must
be a situation in which the student feels a *need*. A need is a feeling
of the organism for something which is absent, the attainment of

which will tend to give satisfaction. The situation is such that the student is motivated to satisfy the need. This creates tensions and drive within the organism which impel it towards its goal. Thus the learner is spurred to physical and mental action, or making a response. The first response often does not lead to the goal; he runs against a barrier. If the motivation to learn is strong enough, the learner seeks another response or series of responses. One after another of these responses may fail to lead to a solution, but finally he selects a path of action that reaches the goal. He has solved the problem; he is ready to readjust his total behavior in this situation. He may go over the solution, to make the meaning and structure more precise, and his formulation more articulate; to make the whole situation more highly differentiated from previous learning, and more generalized, until he has developed a new pattern of behavior that will function in new problems containing the same or similar situations. He has learned.

Each of the several psychologies of learning has its own explanation of the way the learning goes from need to goal. That these theories do conflict at a number of points should not concern us too much. If the application of one of the conflicting theories proves more useful for our purpose in a given situation, and application of another theory in another, we shall use each as it fits the occasion. Until psychology develops into a more significantly unified, scientific theory we must do this. Physicists do this in the the study of light where both the wave theory and corpuscular theory are applied, giving consistent results in some instances and contradictory results in others. Finally the areas in which all of the theories are in agreement will be especially important for our study of learning. Here we shall briefly examine what three theories, association, conditioning, and field psychologies say about learning.

END PRODUCTS OF LEARNING

In the learning of mathematics, a student is expected to do everything from handling concrete objects in counting to making abstract logical deductions with the use of symbols. There exists a sort of hierarchy of end products of learning in which we strive

THEORIES OF LEARNING

for the highest level. One of the simplest types of human learning is a *sensory-motor skill*. The response is practically automatic once it is learned. This is illustrated in teaching a child how to use a pair of compasses to draw a circle.

On a slightly higher level we have *perceptual-motor skill* learning. This can be characterized as learning which is applied immediately to a perceptual pattern. It is illustrated in the learning to use a protractor to measure an angle and a ruler to measure a line segment, or in drawing a geometric figure.

The next type of learning which occupies a large part of the school activities is *mental association*. This is the type of learning which gives the child his store of number facts, names of algebraic terms such as exponent, coefficient, binomial, the names of geometric figures. It includes vocabulary learning.

While a student may learn to recognize an exponent, coefficient, or median of a triangle, he may not fully comprehend these objects of thought. For this purpose he must *learn concepts*. How concepts are learned in mathematics is the topic of an entire later chapter. When a child has a mental image of a thing and can relate it to other things through definitions, laws of operation, application, or generalizations, he does a great deal more than mere identification through association.

A final end product of learning, of concern to mathematics teachers, is *problem-solving* as illustrated in the example in a previous section. Here all of the other end products are brought to bear in making hypotheses, judgments, organizing evidence to give solutions, and forming structures of knowledge such as pure mathematics. The chapter in this book on problem-solving is particularly concerned with aiding the mathematics classroom teacher to develop this type of learning on the part of his students.

It should be noted that while the end products appear quite distinct in form, yet their learning has in common the elements in the learning diagram. In learning to use a pair of compasses for example, there must be motivation, there will be motor movements which will not give the desired circle, then a correct use of the hand and fingers comes forth, finally the learner will try this successful technique until the variability has been reduced to a desired level of manipulative skill.

ANIMAL LEARNING

Conditioning as a theory of learning grew out of laboratory studies of animal behavior. In most of these experiments, the animals were restricted or constrained so as to be unable to avoid stimuli. The reward or punishment was in most cases the same, an electric shock or food, respectively. Since for most animals food is a strong incentive, it was easy to condition the animals and to have them react in a given manner to a given sign as a stimulus. The animal thereafter did what he had been stimulated to do and it was said that the animal had learned.

Cole and Bruce (1) characterized two levels of freedom in animal learning, (a) when the animal is almost totally restrained in a harness and free to move only one or more legs, and (b) when the animal is confined in a cage or maze but free to move about within it. In the first case the animal learns by responding to a stimulus; in the second by selecting from random activity those responses that lead to satisfaction. In neither case can the animal *explain* why he behaves the way he does.

On a higher level of animal learning, Köhler (9) described how, confined in a cage, an ape could piece together two sticks inside his cage, and reach outside the cage to scrape food to within reach of his arm. The ape had previous experience with using a stick as a scraper, but not with putting two sticks together. The ape learned the latter by accident or by random error, but having learned it, he had a flash insight as to its use in getting food.

Animal trainers use the method of conditioning in training their subjects, using a lash and food for punishment and reward. Even fleas and worms can be shocked into behaving as we would have them behave. The questions for the teacher are: Shall we use the techniques of the animal trainer in our classes, imposing the necessary restraints, with accompanying punishment and reward for failing or successful responses? Or shall we permit freedom of learning experience? Or are both techniques of value depending on the time and the nature of the learning.[2]

[2] See film "Willy and the Mouse," 16 mm., 11 min., black and white. Teaching Films Custodians, 25 West 43 Street, New York City.

In the following description of theories of learning, it will be well to recall at all times the limitation of applying animal learning theory to the learning of mathematics. Thorndike, and many others since, in their experiments on animals, conclude that they do not learn by reasoning or by social imitation. They learn *only* by physical doing. They do not develop a culture, or transmit their culture from generation to generation. They do not speak, they do not use symbols, and they communicate only by signs, where by sign is meant the stimulus for a fixed response. In contrast, mathematics is learned by reasoning, by the use of symbols, and by the transmission of cultural patterns. We recognize that the learning of infants by reproof and reward is the same as the conditioning of animals. We also recognize that much that we learn in the early stages of mathematics is learned by doing, and to this end we should examine what conditioning theory has to offer.

CONDITIONING

To enter into a detailed discussion of the theory of the various schools of psychology is beyond the purpose of this book. The interested teacher can obtain this by studying the literature in the bibliography appended to this and successive chapters. We shall state the main principles and characteristics of each of the psychologies and illustrate these with applications to mathematics learning.

Conditioning, with its emphasis on stimulus and response, was one of the first psychological theories carried over to human learning and still either consciously or unconsciously guides the teaching patterns in many of our classes. Since we cannot tell the difference, by examining the brain and nervous system, between a boy who gives a correct response to a problem and one who does not, we resort to predicting from observing what each boy does (his responses) and the situation that brings about his responses (his stimuli). Evidently the boy with an incorrect response is in a situation with unfavorable stimuli, which is different from the situation or stimuli acting on the boy who gives the correct response. If the outside situations are the same, then we can predict the inner (inside the organism) stimuli are different.

The fundamental principle of conditioning as given by Guthrie

(13:23) is: *a stimulus pattern that is acting at the time of a response, will, if it recurs, tend to produce the same response.* According to this theory we learn only what we do in a given situation. We learn only our reactions, our responses. To bring about learning we, as teachers, must induce our students to follow certain mathematical patterns or behaviors, at the same time that they are confronted with stimuli. These stimuli become the signals for the mathematical behavior, and when this has been done, the signal replaces the inducement, that is, the response tends to come forth with the signal thereafter and learning has taken place.

Thus the learning of the addition facts, under this theory, can be brought about by having children combine groups, for example placing 3 chairs and 4 chairs in a row. At the same time the children are confronted with the stimulus $3 + 4$ and the response 7 is given. Thereafter any stimulus pattern similar to $3 + 4$ will tend to evoke the same response 7. Thus the child is learning what he is doing at any given time.

We can learn incorrect responses as well as correct ones, and in this case it is necessary to break down the incorrect response. If an algebra student says $(a + b)^2 = a^2 + b^2$, we must remedy the situation. To do this, conditionists use *associative inhibition* in the following manner: present the signal $(a + b)^2$ and along with it a stimulus for inhibiting the incorrect response (a teacher's disapproval—no, no; or the correct response, $a^2 + 2ab + b^2$ are all inhibitory stimuli). After sufficient repetition the incorrect response will be forgotten. In this case forgetting is failure to respond to a signal and it is due to new associations formed, that is, learning to do something else that is more desirable.

In general, conditioning has as its basis for learning:

1. The making and breaking of habits, the acquisition of skills.
2. The response to a pattern of stimuli is conditioned. We learn what we are doing. We learn incorrect responses (errors) as well as correct responses.
3. New responses result from conflict and inhibitory stimuli.
4. Learning occurs normally in one conditional response. The need for repetition in skill learning is that a skill is not simple, but it is a large collection of habits.
5. Learning best takes place when a desired response is asso-

ciated with appropriate signs, gestures, mathematical symbols, and words that act as stimuli for the desired action.

6. Since we learn what we do, we must be free to act. Hence learning takes place best in a free situation, not in a forced and harnessed activity.

It is easy to see under these principles how rote learning and fact learning can be brought about. It is rather difficult to see how these principles aid in learning to solve an original problem in geometry or algebra. For this latter purpose the theory takes recourse to trial and error which is explained better under connectionism.

CONNECTIONISM

The fundamental characteristic of connectionism is the bond established between a situation and the response made by the organism. These bonds become unified and patterned through selection (trial and error) according to certain laws of effect, exercise, readiness, and analysis.[3] The degree of learning to which the organism can aspire is largely determined by its inherited qualities. As the organism matures, it develops connections (habits and skills) which must be practiced to achieve permanence. The more complex the acquired bonds can become, the greater is the capacity to learn mathematics.

The law of effect has particular interest for learning mathematics. It says: *A bond is strengthened or weakened according as satisfaction or annoyance attends its exercises, and reward upon success is the most potent factor for insuring learning.* If this is so, we learn, practice, and have an interest in those things which are pleasurable. Thus, the first experiences a pupil has in mathematics should be simple enough to insure successful results and should be accompanied by reward in the form of praise or encouragement. *Start right, and practice.* Under connectionism some adherents say it would be detrimental to learning to allow a student to flounder about or to make mistakes. Others, patterning their belief on

[3] These laws are stated by Sandiford (13:111). While in the early formation of his theory of learning, Thorndike states these as specific laws. He and his followers later amended these statements to serve as descriptions or characteristics of learning rather than as laws. They should be thought of in this latter aspect in this book.

animal experiments, say the initial experience should be entirely free to permit a choice of any trial, no matter how blind. Out of this experience should come direction of later efforts. When a correct bond has been established, as for example $a^n \cdot a^m = a^{n+m}$, immediate and frequent repetition will strengthen the bond and give greater probability of its functioning in a similar educational situation at a later time. So we should drill repeatedly on the thousands of facts throughout the sequential learning of mathematics.

The law of exercise says: *When a connection is made between a situation and a response, the strength of the bond is increased; when the connection is not made over a period of time the strength of the bond is decreased.* Thus in learning to solve a quadratic equation by the use of the formula, the oftener the equation is accompanied by the proper use of the formula, the stronger the bond, and in later appearances of quadratic equations, the formula is more apt to come as a response. Further, the sequence of operations in applying the formula—equating the function to zero, determining the coefficients, substituting, simplifying the result, checking—form a *belonging sequence*, the repetition of which accompanied by success or other reward promotes learning.

The law of readiness says: *When a bond is ready to act, to act gives satisfaction, not to act gives annoyance. When a bond is not ready to act, and is made to act, annoyance is caused.* Thus to attempt to make a child form addition facts or learn the multiplication table, when his organism is not ready to act, is to cause dislike, and to interfere with later learning of the arithmetic. If a child cannot substitute 10 pennies for a dime in a practical subtraction of 23 cents minus 15 cents, then he is not ready to do subtraction involving borrowing (changing a ten to ten ones) and to force him to do the abstraction would interfere rather than aid his later learning of the process.

The law of analysis (similarity and dissimilarity) says: *When a given response has been connected with many different situations which differ in all respects except one common element, the response becomes bound to that element.* In later situations totally different from the previous situations, the presence of this common element will tend to evoke the given response. This law is closely related to trial and error learning or problem-solving. A child is confronted with three blocks and one block put into a single group, and hears or gives the response *four*. Then two apples and two apples, with the same response; then various patterns of four objects of various kinds, all with the same responses. In all cases the situation is different except for the fourness. The law of analysis says that in any later complex situation the recurrence of fourness will evoke the response four. This is how a child learns *four*. Thus analysis, in the sense used here, fosters learning.

Under analysis, trial and error is not a befuddled, blind chance affair. On a given trial we do not get a desired response so we discard the mental path we used and select another path to our goal. We do not return to unsuccessful paths (errors). Thus trial and error is deliberate choice. Each succeeding trial takes less time until finally we solve the problem. A permanent bond is then formed between the stimulus and the goal and in later different situations in which the stimulus occurs (along with many other elements) this goal response will come forth. Thus the $S \to R$, $a^2 - b^2 \to (a - b)(a + b)$, as a desired learning should be taught in many different situations involving $a^2 - b^2$ as a common element but always with the same response $(a - b)(a + b)$. Then when the right triangle occurs with hypotenuse r and one side x, the situation $r^2 - x^2$ should evoke $\sqrt{(r - x)(r + x)}$ as the remaining side.

The law of analysis indicates that all complex learnings should be analyzed into simple elements, and then taught or presented in a sound, pedagogical order. If you wish to learn how to add a/b to c/d, analyze every step involved—the definition of a fraction, changing a fraction to an equivalent fraction, defining a common denominator, finding common denominators, changing fractions to common denominators, the rule of adding numerators —then drill on the process until it is mastered. This is the way much of our mathematics is to be learned.

The recent war and the present mobilization are focusing our attention on knowledge as a tool. The goal of learning is performance. To this end we have stressed the learning of facts and skills and connectionism has been the prevailing psychology. Under this theory our whole program in mathematics has been largely concerned in getting students *to do* their operations quickly and accurately whenever they occur. Problem-solving in mathematics is reduced to a method of steps to be followed in a belonging sequence, from step one (read the problem) to step n (check in the original problem) which, if mastered, will automatically lead to the solution of the problem. Drill is the keynote to achievement. If the multiplicity of facts and skills becomes so great that it finally overwhelms the learner, he has reached the limit of his inherited capacity. With sufficient capacity one can learn all the mathematics as a set of sequentially ordered and related facts.

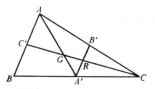

The following geometrical original may be used to illustrate how connectionism leads to a solution. In the triangle shown, AA' and CC' are medians. (These are stimuli which evoke the response A' and C' are mid-points of the sides.) AA' and CC' meet at G. (This evokes the response that AG is $\frac{2}{3}AA'$ or GA' is $\frac{1}{3}AA'$.) B' is the mid-point of AC. $A'B'$ meets CC' at R. What part of $\triangle ABC$ is the $\triangle A'GR$?

The student begins his goal-seeking by trial and error: $\triangle A'GR$ is a part of $\triangle GA'C$ which in turn is a part of $\triangle AA'C$. But $\triangle AA'C$ is $\frac{1}{2}$ of $\triangle ABC$. (This is a transfer of an identical element in many previous problems in geometry.) What relation has $\triangle GA'C$ to $\triangle AA'C$? Here the response $A'G$ is $\frac{1}{3}AA'$ may be forthcoming (and if it is not, the student may be given a cue that suggests this previous learning). Then $\triangle A'GC$ is $\frac{1}{3}$ of $\triangle AA'C$ and thus it is $\frac{1}{6}$ of $\triangle ABC$. (The student evokes the conditional response $\frac{1}{2}$ of $\frac{1}{3}$ is $\frac{1}{6}$.) Now how can I find what part $\triangle A'GR$ is of $\triangle GA'C$? Since vertex A' is common, he thinks of bases GR and GC. This evokes drawing an altitude from A', but this seems to complicate the figure and this trial is rejected. Finally the student says, "I can't find any relation." The relation of GR to GC and RC does not come forth because he has never had this response in his past learning. The goal is at his door but he fails to make the connection.

What usually happens is the abandonment of this procedure and a search for another way. He says, $\triangle AC'G$ is also $\frac{1}{6}\triangle ABC$. Are $\triangle A'GR$ and $\triangle AC'G$ related? $A'B'$ is parallel to AB. (The past learning of a line joining the mid-points of two sides of a triangle evokes this response.) Immediately the pattern of similar triangles $AC'G$ and GRA' is recalled since this is a very common learning. Similar triangles can bring forth many responses but the common element here is area and this recalls the fact that areas of similar triangles are in proportion to the square of their corresponding sides. Side GA corresponds to GA' and their ratio is 2 to 1. Squares bring forth 4 to 1. Then $A'GR$ is $\frac{1}{4}\triangle AC'G$ which is $\frac{1}{6}\triangle ABC$. The numbers $\frac{1}{4}$ of $\frac{1}{6}$ finally give $\frac{1}{24}$ as the solution. Under connectionism, the answer $\frac{1}{24}$ would be given at the beginning of the problem, and it is the path from the given to the conclusion that is sought, not the discovery of the relation $\frac{1}{24}$.

Of course, in the above, other unsuccessful trials may have been made and then discarded, until the path to the goal is made. The student then repeats his solution several times, each attempt taking less time until he has made the solution readily available for further use in future learning situations.

The reader, no doubt, can supply many similar examples from arithmetic, algebra, trigonometry, or mathematical analysis. The principal characteristics of this theory of learning are:

1. Thinking back to similar situations to find a particular response that worked previously; the transfer of identical elements.

2. Trial and error, discarding unsuccessful paths (responses); avoiding wrong responses.

3. Each complex situation is to be broken up into a series of simple elements arranged in a sequential order. Each simple element is mastered separately. The seriated set of mastered elements make up the whole.

4. After the whole solution is obtained, repeat and drill until the solution is sufficiently strengthened (conditioned) for later recall.

5. Reward successful learning of desired goals.

It should not be assumed from the foregoing discussion that connectionism was not concerned with organized systems of related knowledge. Quite the contrary, Thorndike consistently insisted on organization and interrelatedness in learning. He said:

"Arithmetic consists not of isolated, unrelated facts, but of parts of a total system, each part of which may help to knowledge of other parts, if it is learned properly. . . . *Time spent in understanding facts and thinking about them is almost saved doubly*" (17).

"Knowing should be not a multitude of isolated connections, but well ordered groups of connections, related to each other in useful ways . . . a well ordered system whose inner relationships correspond to those of the real world . . ." (18).

"Every bond formed should be formed with due consideration of every other bond that has been or will be formed; every ability should be practiced in the most effective possible relations with other abilities" (19).

If connectionism is held to be not adequate, it is not for its objectives, but in its means used to secure the objectives. Through its emphasis on the detailed analyses of every mathematical process into a large number of serially related bonds to be practiced, the ultimate outcome of fundamental concepts, generalizations, and organizations frequently failed to materialize. "The forest could not be seen because of the trees." Recently psychologists are developing a more adequate explanation of learning through which understanding and well-organized patterns of knowledge come to the fore. The results of their study are now considered.

FIELD THEORIES

A major desired outcome of school education is an ability to solve problems. We are determined that this ability will be permanent and grow stronger. While we learn many facts and skills, it is the developing of the process by which they were learned that is as important as the material learned, for it is this procedure that will enable us to "go learning," to solve new problems. The goal is thus *to learn how to learn*.

It is in this aspect that field theories differ most from other theories of learning. In conditioning, the ceiling of learning is dictated by the inherited capacities of the organism. In gestalt theory, the inherited capacity is increased (modified) within limits by training. There is a body of mathematical knowledge that, regardless of the capacity of the learner, could never be acquired without appropriate prior physical symbolic and linguistic experience. When this experience is acquired the learning ability rises. Thus experiencing and vocabulary building can increase the inherited capacity of the organism to learn, and they become important elements of the learning process.

Experiencing a situation and finally understanding the situation calls for a study of the *whole* of it, rather than a detailed study of the individual elements of the situation. It is only as the relation of a part to the whole is sensed that a solution of a problem can emerge that will be permanent. This is one of the fundamental principles of field theory: *Always consider the whole situation in responding*. It is not how many facts you know about a situation (a geometric original or a problem in installment buying) but how much *relatedness* in all possible ways there is between the facts and the whole of a situation. For example, a median is not only a bisector of the opposite side of a triangle, but of every line segment in the triangle parallel to this side.

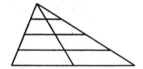

When the various elements of a situation are grasped in their relation to the whole situation, *insight* occurs. Insight is thus the final outcome (behavior) of a given situation. Before insight leading to the solution of a problem occurs, there may be trial and error, but it is not the type explained by connectionism which rejects false leads, but rather an *analysis* of the relationship of parts to the whole and the seeking of those relationships which give the complete understanding of the problem. According to connectionism the rejection of each false lead (error) brings one closer to the level of his goal; according to gestalt each analysis helps, but there is not necessarily a closer approximation after each analysis, only a larger number of relations. When all the relations discovered shape into an organized pattern, there is insight and the problem is solved. The following diagrams illustrate these two points of view:

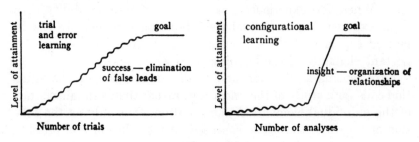

EXAMPLES

A study of two problems may aid in clarifying the concept of total configuration and insight. Consider the geometry original given on page 16. The problem is, "What relation has the area of $\triangle GA'R$ to the whole $\triangle ABC$?" A gestaltist would not give the answer $\frac{1}{24}$, but expect the student to discover it. According to

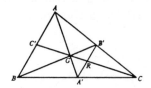

gestalt, the first emphasis is to be on the whole figure, and its related parts. We are to study the relationship of the whole to its parts, and of the parts to the whole. Looking at the figure this way, we draw the remaining median BB' and note:

$\triangle AGB = \triangle AGC = \triangle BGC = \frac{1}{3}\triangle ABC$

$\triangle B'GA' \sim \triangle AGB$ (They look similar, and $A'B' \parallel AB$)

$\triangle B'GA' = \frac{1}{4}\triangle AGB = \frac{1}{12}\triangle ABC$

$\triangle GRA' = \triangle GRB' = \frac{1}{2}\triangle GA'B' = \frac{1}{24}\triangle ABC$.

As soon as the pattern $\triangle GRA' = \frac{1}{8}\triangle AGB = \frac{1}{24}\triangle ABC$ emerges (or a similar pattern), insight has occurred.

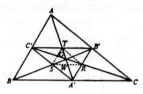

Another approach is to draw all the medians and then analyze the figure. In this case the parts are in closer relation (in a narrower field) than in the above figure, and almost by flash you see
$$\triangle A'B'C' = \tfrac{1}{4}\triangle ABC; \quad \triangle A'GR = \tfrac{1}{6}\triangle A'B'C'.$$
It is also significant that the latter figure has within it elements of a new learning situation, which was not apparent in the other figures. Real creativeness lies not so much in the solution to a problem, as it does in the significant new problems that emerge out of a solution. This is in part the element of *generalization* which gestalt psychologists hold essential for permanency of learning and transference to new situations. In the figure, the relationships evolved suggest drawing SR, ST, TR, and continuing on in this manner. Then $\triangle GWR$ has the same relationship to $\triangle A'B'C'$ as $\triangle GRA'$ had to $\triangle ABC$. This immediately suggests two infinite series of areas, 1, ¼, ⅟₁₆, ⅟₆₄, · · · and ⅙, ⅟₂₄, ⅟₉₆, ⅟₃₈₄, · · · . This aspect of *discovery* and *extension by generalization* is a major aspect of configurational learning.

Finding the rate of interest charged on an installment loan illustrates several of the types of learning. Consider the problem: A television set can be bought for $300 cash, or for $60 down and 6 monthly payments of $45. What is the approximate rate of interest paid on the installment loan? At times a formula is given and the students substitute. They are conditioned to do this by the recall of the formula, say

$$= \frac{2fI}{P(n+1)}, \qquad i = \frac{2(12)(30)}{240(6+1)} = 42.86\%,$$

and the knowledge of what each letter represents. They substitute numerical quantities, carry out the operations and give the result. Thus the desired end response of learning (a formula) is *given* with certain stimuli and the association is made. It is then practiced in sufficiently different situations to secure its desired strength.

Those who believe in conditioning would analyze the problem into its separate parts and arrange them in a belonging sequence, perhaps (a) finding the total installment price, (b) finding the cost of the installment plan, (c) finding the amount borrowed, (d) finding the size of each monthly principal payment, (e) finding the total time of the loan, (f) finding the interest rate. Each of these elements would be practiced separately until known, and then in the order given until known. A student who had finally learned would make the proper series of related stimulus-response actions and get the answer. His actions would probably be as follows: (a) $60 + 6 \times 45 = \$330$, the total cost; (b) $\$330 - \$300 = \$30$, the cost of the plan; (c) $\$300 - \$60 = \$240$ the principal borrowed; (d) $\frac{240}{6} = \$40$, each principal payment; (e) $\frac{1+2+3+4+5+6}{12}$ = $1\frac{3}{4}$ years, the time of the loan; (f) $30 = 40 \times i \times \frac{7}{4}$ or $i = 42.86\%$, the interest rate.

In field theory, the words configuration or gestalt (form) are constantly used. For this reason, the psychology using field theory is referred to as *gestalt*. A configuration is a pattern of all the elements entering into the perceptual field of the learner. If the elements of the field are reorganized, a new pattern or configuration is formed. While the elements may be abstract, it was frequently found helpful to represent the configuration by a geometric form, in which each of the geometric elements symbolizes an element of the field, and the positions of the geometric elements indicate the relationships of the field elements to each other and to the total situation. More generally, however, a gestalt is to be considered the total situation with which the learner is confronted.

The field theorist would try at the start to bring all the elements of the problem together. He would not hesitate to use a geometric drawing to aid the mental organization of the elements. The following figure (derived with the help of the students) is the initial step. (Several other configurations are possible.)

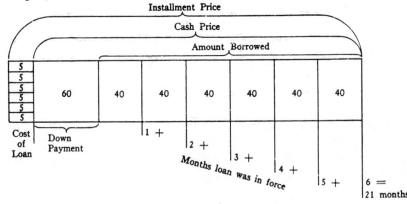

A study of the diagram allows a simpler rephrasing of the problem. If $30 is paid on a loan of $40 for 21 months, what is the interest rate? A student may now recall the simple interest formula and solve the problem. Having done this problem, and several like it, he would then generalize (pattern, structure, organize) his solution and obtain as the end result, the formula. The formula is the result of his learning, not the starting point.

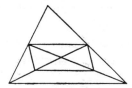

FURTHER ASPECTS OF FIELD THEORY

When a field theorist says "The whole is greater than the sum of the parts," he is not referring to the physical characteristics of a situation, but the concepts and relationships involved. One can learn (a) a median is . . . (b) a line joining the mid-point of two sides of a triangle is . . . (c) a parallelogram is . . . (d) diagonals of a parallelogram . . . as four relationships. Now put them together in a whole and we also have (e) medians meet two-thirds. . . . Thus any whole learning has within it concepts and potentialities for further learning that are greater than the sum of each of the elements that make up the whole.

Further, it is easy to recognize that if learning is to be extended, each whole is only a part of a greater whole. By total configuration, then, is meant all the elements that come within the perception of a given situation. Learning is the integration and reorganization of the elements of a given situation into a mental pattern. When the pattern is finally organized, it is done swiftly, in a flash. The rest is a matter of progressive clarification or smoothing the performance, which is the *function of drill or practice*.

In summary, the various studies in field psychology explain learning by the following characteristics:

1. Initial learnings come from experience (physical and mental experiments), constructive methods, not from definitions. It is the dynamic aspects of events that aid learning. Whatever is to be learned must have its roots in some challenging, problem-presenting situation.

2. All parts related to the learning situation must be brought into focus to see the problem as a whole. Scattered elements or isolated details prevent insight.

3. The analysis and obtaining of relations of parts to whole and whole to parts, the recalling of past patterns of learning, and blending of the given elements permit the restructuring of the elements into a new pattern. When this occurs the student has *insight*. It is here that abstractions and generalizations come to the fore. It is the analysis and insight that give meaning to arithmetic, algebra and geometry. Lack of variety of previous experience, and over-preoccupation with fixation of specific habits, operate to prevent insight.

4. After insight, the student practices the solution to smooth and clarify the new learning (structure). The more sharpened and systematized the knowledge is, the less chance is there for forgetting.

5. A whole (configuration) is always a part of a greater whole. The relationships in one configuration (e.g. congruent triangles) appear and are generalized in later configurations (e.g. similar triangles). The relationship of relations is organized into a structure of knowledge through analysis, synthesis, and deductive logic. A system of knowledge must be built. We draw from the system (and not from a multiplicity of isolated facts) for further learning. Thus project learning and systematic courses are not contradictory but of a different level of maturity in the learning process.

SOME EXAMPLES OF MATHEMATICS LEARNING

There are several methods of teaching the multiplication facts. One common procedure is to present the facts, simpler facts first, and drill on responses until they are automatic and correct. Thus the learner is conditioned, or makes the connection-stimulus 6×9 —response 54. Then the pupil is taught how to apply this to solving problems. He goes from given specific facts to experience. Another procedure less common is to have pupils build their own facts out of experience, and then organize them into related tables. Which of these is the best learning procedure?

THEORIES OF LEARNING

Multiplication

	0	1	2	3	4
1	0	1	2	3	4
2	0	2	4	1	3
3	0	3	1	4	2
4	0	4	3	2	1

Consider the table shown of 5 numbers only, and call it a multiplication table. Make flash cards to show the items to be mastered: e.g., $2 \times 3 = 1$, $3 \times 2 = 1$; $4 \times 3 = 2$, etc. Now drill. The reader can assure himself that this table of multiplication facts can be learned in five minutes. With appropriate drill of five minutes each day for a week it can be made fairly permanent. With continued drill at spaced intervals over a year it can become as permanent as "hickory, dickory, dock." But what does it mean? To most elementary-school teachers, and to many readers, the answer is, "Nothing." It is merely an association of meaningless responses to given stimuli, which can be learned.

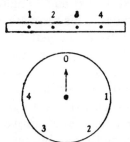

Now consider the dial and its control, as shown in the figure. Each time the button 1 is pressed the dial moves one space, clockwise. It is easy to see that 2 times 1 is 2, 3 times 1 is 3, and 4 times 1 is 4. If the button 2 is pressed, the dial moves through 2 spaces, clockwise. Thus $2 \times 1 = 2$; two moves of 2 spaces is 4, and three moves of 2 spaces places the dial at 1 and hence we say $3 \times 2 = 1$. Similarly, $4 \times 2 = 3$. If the button 3 is pressed 3 times, the dial makes 3 moves, each through 3 spaces and ends at the mark 4, hence 3 × 3 is 4.

From this experience it is easy to build up the table and establish meaning for it. If we are to use the table of dial multiplication efficiently, we had better practice the table to make the results automatic. But we can interpret any result in the table readily. The use of the dial to establish the table and then apply it is the less common way of teaching the usual multiplication facts in school, but it would appear to be a more satisfactory, need-fulfilling method than specific conditioning.

Luchins performed an experiment in solving originals in plane geometry. His subjects were high-school, tenth-grade students. The students were drilled on proving lines and angles equal by using corresponding parts of congruent triangles. They developed

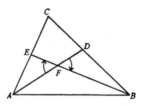

a set or were conditioned so that equal angles evoked the response parts of congruent triangles. This response acted as a stimulus to get sufficient parts of the triangle equal, to establish congruence—then the final response, corresponding parts of congruent triangles are equal. A series of originals on proving angles equal was given to the class after the above learning. In the set, the following original occurred: In isosceles triangle ABC, the bisectors of the base angles, AD and BE, meet at F. Prove $\angle AFE = \angle DFB$. So strong was the conditioning that all but one of the students proved the triangles AFE and BFD were congruent in order to arrive at the equality of the angles, whereas the solution should have been apparent at once by the use of vertical angles. A mind-set procedure (conditioning, transfer of identical elements) was a hindrance to learning.

The following sequence of problems was presented to a large group of teachers of mathematics, all with the same result.

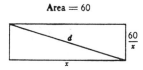

Area = 60

1. Represent the altitude of a rectangle of area 60 as a function of the base. The problem was a simple recall of an area pattern and the answer given was $h = \dfrac{60}{x}$.

2. In the same rectangle, represent the diagonal as a function of the base. Continuing from problem 1, most students responded by

$$d = \sqrt{x^2 + \dfrac{3600}{x^2}} \quad \text{or} \quad d = \dfrac{1}{x}\sqrt{x^4 + 3600}$$

3. In the same rectangle, represent the diagonal as a function of the perimeter. Continuing from problem 2, all students responded by placing $p = 2x + 2\dfrac{(60)}{x}$ or $x^2 - \dfrac{px}{2} + 60 = 0$. They obtained $x = p \pm \dfrac{\sqrt{p^2 - 960}}{4}$ and attempted to substitute this in the expression for d in problem 2 with various results from the involved algebraic manipulation.

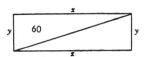

The students were then confronted with the rectangle, each part labeled with a unique symbol, and asked to study the relationship of the parts to the whole. They wrote $x^2 + y^2 = d^2$, $xy = 60$, $(x + y) = \dfrac{p}{2}$, and it was only a very short period before the "ahs" came forth as insight gave the pattern $(x + y)^2 = x^2 + 2xy + y^2$ or $\dfrac{p^2}{4} = d^2 + 120$. A set belonging-sequence can help recall old patterns but it can interfere with the discovery of new patterns.

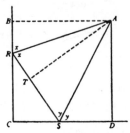

A geometric original that has caused much trouble for high-school teachers as well as their students is the following one.[4] In the figure shown we start with right triangle RSC and bisect the exterior angles at R and S. These bisectors meet at A. From A we draw perpendiculars to the sides RC and AD forming rectangle $ABCD$. Is it a square?

To prove the figure a square we must prove two adjacent sides equal. At once congruent right triangles is the response. An attempt is made to get angles equal through the use of complementary angles and angle sums. This fails. Even the relationship that x is not necessarily equal to y in this general configuration fails to bring out the response, "try something else." Despite the failure, the students continue to return to the unsuccessful path of obtaining congruent triangles. They fail to use the relationship of the given parts to the whole configuration. An approach that studies the whole configuration says "What is the pattern of angle bisectors?" This would suggest perpendiculars AB and AT for bisector AR, and perpendiculars AT and AD for bisector AS, and immediately insight is obtained, $AB = AD$. It is only in rejecting unsuccessful paths that trial and error can lead to a solution. It is in the relation of the whole configuration to its parts that an insightful solution emerges.

When we see an immediate relation between a given condition and a desired goal, that is, when the recall is forthcoming by simple association because of the simplicity of the problem, conditioning seems to be a good explanation for learning. When, however, the situation is complex, and we cannot see the path to the solution, the use of seriated bonds, the study of chain responses through a study of the parts of the situation seem to have much less value than a field approach of studying the whole problem

I am indebted to Miss Barbara Betts of Boston for this example.

and making an analysis of the relationship of the parts to the whole, parts to parts, and whole to parts.

It is in the study of examples of learning and the various theories about learning that a teacher can gain a philosophy regarding a psychology that will work for him in his classroom activity. In the rest of this chapter we shall try to seek some common principles of learning that are concrete and acceptable for classroom management.

AREAS OF AGREEMENT

Most of us, with a little introspection (and perhaps some rationalization) can recall experiencing most of the elements in all these psychologies of learning and the examples cited. In solving a geometric original, how frequently we stabbed at one route, then another, meeting block after block, coming back to the facts we started with, stopping and resting, and then going on until suddenly the solution appeared. Was it trial and error, or progressive clarification? Did we take separate steps in a sequential order or did we analyze the relationships of the parts of the figure to the whole configuration? Was there "insight" as the configuration became clear? Did each step we took put us in a frame of reference that made us take the next step? Perhaps all of these to some extent.

If you offer a mechanical puzzle to a class of students and they are not too homogeneous, you can observe a hierarchy of learning situations. Some students will merely shake, pull, and shove the puzzle by almost blind trial and error (no thinking) and by sheer accident they may solve the puzzle. Asked to try again they do the same stunt for hours and do not solve it. Others will attempt a deliberated trial and error procedure, remembering false leads and avoiding them in subsequent trials. Eventually a series of selected trials (belonging sequence) gives the solution, and then repetition of the successful sequence insures permanency of the solution. Still others will study the whole mechanism and the relationship of the parts to each other and the whole puzzle before any attempt at solution. They will try a certain movement, not necessarily to reject this movement if it does not succeed in solving the puzzle, but to see how it is related to other parts. After study, experience and trial with errors, the solution is obtained as an organized pattern. There is little need to practice each operation, but there is need to have the whole puzzle pattern clarified and verified.

The classroom teacher is desirous of having at his command a fundamental set of acceptable principles of learning upon which to organize his teaching so that the best possible learning takes place. At first glance the several interpretations of learning seem so different and in some respects contradictory, that confusion seems the outcome. But on further consideration, the teacher will find certain elements in each of the theories that appear to be important to him, and many other elements common to all the theories. We can derive from all the psychologies a theory of learning that is effective for ourselves and modify it as we ourselves learn more. The following elements may serve as a foundation of an effective theory of learning.

1. There must be a *goal* on the part of the student to learn. The learner must be aware of this goal. Thus a teacher must not only know *why* a student should learn to solve a quadratic equation, but he must know how to transform this *why* into a recognized goal on the student's part. Motivation conditions the quality of the learning. A pupil will stop counting and learn addition facts when counting becomes inadequate for him and he desires a more efficient method.

2. All cognitive learning involves *association*. The situation-response may be simple or complex, it may be patterned, but it is an important aspect of learning. When we see a^3 we expect the response $a \cdot a \cdot a$. Even a relationship of one element to another is a form of association. The situation, similar triangles, is expected to bring forth the response—proportional sides and equal angles.

3. We recognize trial and error or analysis in most learning. If it is blind groping, then the learning situation is bad and the learner is very immature. If the trial and error is deliberate, then it can be better called approximation and correction, or analysis of relations, which continues until insight occurs. The learner should not be allowed to flounder. He should be guided in his experimentation and activity toward the final goal.

4. Learning is complete to the extent to which the relation-

ships and their implications have been understood. These relationships sometimes are learned on an initial trial, especially if attended by an emotional response. More usually, practice is needed, that is, more study of the situation. This may be accompanied in some situations by studying the related set of responses, and in others by analyzing the whole situation. The more simple the pattern to be evolved, the less analysis is needed.

5. The learner must be in action, mentally and/or physically. In conditioning, the learner learns what he is doing. In connectionism, the learner must react correctly to a mathematical stimulus. In field psychology the learner experiments and organizes a pattern. It may or may not be a correct pattern of knowledge. Unless the learner is active mentally and physically, and his actions lead to success, he is not learning.

6. Intrinsic reward of success and awareness of progress toward a goal strengthens the learning and the motivation for further learning. Punishment is a deterrent rather than an aid to learning. Praise a successful response. Encourage students to make a new and different response when their first response is incorrect. With success, students raise their level of aspiration as well as their ability to solve new situations.

7. Discrimination of attributes (abstraction) and generalization are essential to effective learning. Thus all learning situations should be of the type where a relationship can be abstracted and a process can be generalized. This is only possible if the situation is meaningful.

8. New learning is in part a matter of transference of past learning. The degree to which this takes place depends on the degree of similarity of the new situation to the original learning situation, the learner's ability to analyze relationships, and the amount of varied experience in previous learning.

9. We learn facts and skills and we also learn how to learn. Our learning situations might well be changed from "topics" such as factoring, parallel lines, law of sines, and similar things, to problem situations involving the material to be learned.

10. We also learn feelings (attitudes). From unsuccessful experiences we learn to dislike mathematics and to shun the subject. We also learn to dislike teachers of subjects in which we have unfor-

tunate experiences. We also learn to like mathematics from happy experience with it. There are many concomitant learnings that accompany the mathematics lesson that are outside the actual subject matter.

INTROSPECTION AS A GUIDE TO LEARNING

The task of our secondary schools is to establish within the minds of our students the fundamental bases for productive thinking. Unless we have left our students in a position where they can solve new problems—where they can go on with their learning independent of the teacher—we have accomplished but little in our mathematics instruction. The work of the teacher is to develop learning ability.

How we learn has been of interest to others besides psychologists. From the time of Plato, philosophers and educators have attempted to explain how we think. They have attempted this explanation through an introspection of how they themselves and how others have come to know whatever they do know, and to act however they behave. While this may appear to be an unscientific approach, yet the results of the thinking of these philosophers have had wide influence in establishing learning procedure.

Most famous of modern interpretations is John Dewey's *How We Think* (3) written in 1910. Dewey's interpretation of a complete act of thought (the solution of a problem) consists of five major phases: *problem-presenting situations, analysis, hypothesis, deduction, verification*. Each of these areas can be related back to some one or several of the psychological aspects of learning, and to do this can aid us as teachers in establishing our credo of learning.

A problem-presenting situation, or dissatisfaction, occurs when an individual is in a situation in which he is confused, or in which his previous knowledge does not give him satisfaction; he is not adjusted. The individual's previous ways of acting in a situation are inadequate. A student in setting up a problem obtains a quadratic equation which he does not know how to solve since his previous experience has been only with linear equations. He does not know what to do. He may experiment or flounder about, but he is dissatisfied and unhappy. We recognize here the concept of felt need, and of goal-seeking behavior, since no past learning exists to give immediate satisfaction. We also recognize that learning begins in a concrete problematic situation in which the answer is desired by, but unknown to, the individual.

A dissatisfaction causes the student to make a diagnosis of the situation, but only if the motivation is strong enough. If by trial and error, a student can find an acceptable answer to his quadratic equation, the puzzlement may cease, but a further analysis of the quadratic situation will not go on. New learnings are not desired when old ones suffice. We must make the old way of acting so inadequate that the motivation for new knowledge becomes strong enough to send the student on. A quadratic equation with no rational roots may do this.

The analysis is an examination, within the mind of the student, of the situation in which there is dissatisfaction. He discovers why he is dissatisfied and clarifies the goal that would give satisfaction. He recognizes or states his problem. The girl who, upon solving the quadratic $y^2 - 3y + 4 = 0$, obtained the answer $\frac{1}{2}(3 \pm \sqrt{-7})$ was in a state of perplexity because of $\sqrt{-7}$. "It just couldn't be," so she said, "for the number under the radical sign should be positive." But it wasn't positive, and it had to be accounted for. Investigating past experiences in mathematics of creating negative numbers and irrational numbers, which was an analysis or diagnosis of the situation, she finally stated the problem: "I shall have to find an interpretation for $\sqrt{-7}$ to make it meaningful as an answer."

In some theories of learning little is said about the awareness of the problem. Most theories assume the existence of a problem. Thus Dewey has shown that analysis is used not only in the solution of the problem, but much earlier in the study of the difficulty, in the clarification of the desired goal. Unless the learner can detect his goal as a verbalized or unverbalized expression, he will flounder in his learning of mathematics. Thus a student meeting the quadratic equation for the first time should analyze his problem as "I must learn how to find the root of the equation when an x^2 is in it."

The third element of thinking (learning) is a search for hunches, promising leads, tentative hypotheses. This is related directly to trial and error, making responses, or analyzing relations. Testing a hypothesis is making a trial. Recalling the solution of linear equations, a student may try putting x terms on one side of a quadratic equation, and the constant on the other. If the equation is $x^2 = 25$, the hypothesis may prove fruitful, but if the equation is $x^2 + x = 25$, the trial fails (the hypothesis is not sustained). Then a new response is made. If the learner is active and purposeful, he may go back and make a further diagnosis to differentiate between the two types of quadratic equations. In forming hypotheses, the learner may need the help of "cues" to recall those past patterns of learning that will help. It may be suggested that the equation can be written in the form $x^2 - 4x + 3 = 0$ and the student asked to focus attention, for the time being, on the left-hand member (this is not to ignore the right-hand member). In what situations has he seen such expressions before? This may recall the factoring of a trinomial into two linear factors. Here is a recall of identical elements (learned patterns) and perhaps with it the response $x^2 - 4x + 3 = (x - 3)(x - 1)$. This pattern suggests two linear equations. What is the relation in the whole form $(x - 3)(x - 1) = 0$ between the left-hand member and the right-hand member? Another hypothesis (or trial) is, "Let each factor be zero." This hypothesis gives $x = 3$ or $x = 1$, and either answer satisfies. The student has broken the block and reached his goal. Note the importance of the recall of similar patterns (association). In framing hypotheses, an important question is, "What have I learned before that can be of help in solving this new situation?" Students should be imbued with this question.

The whole process of framing and testing hypotheses until a satisfactory route to the goal has been reached is the heart of the learning process. It is the most difficult part, and if a student is not successful after a few trials, he may deem the problem too difficult to solve. Many of the readers can recall "giving up" in their own learning, and allowing the problem to rest until more elementary forms were mastered, or until a further diagnosis of the problem could be made. A situation can be too difficult to be mastered, or the student may be too immature (not ready) for the task. It is important to note that the student must not be told how to solve the equation. He must be guided to make his own analyses, his own hypotheses, his own trials, and arrive at his own final solution. The more he does this, the better learner he will become. We recognize in this much of the field psychology point of view on learning.

The fourth element in Dewey's analysis of learning is deduction. By this is meant the organization of the solution of the problem into a logical frame of reference. To Dewey, this was the most important phase of the learning process, and most of his book, *How We Think* (3), is concerned with logical organization. He says, "Information ... is not merely amassed and then left in a heap; it is classified and sub-divided so as to be available as it is needed" (p. 41). And further, "Only deduction brings out and emphasizes consecutive relationships, and only when *relationships* are held in view does learning become more than a miscellaneous scrap-bag" (p. 97).

Once insight has occurred, and the goal has been reached, it is seldom the case that the whole pattern is so clear and distinct that it will be recalled when needed in later learning. The organization of the pattern into a logical construct, after insight has occurred, can come about in several ways, (a) by going over and then generalizing the particular solution, (b) by taking many similar examples and abstracting the common elements of solution, (c) by making a logical chain of known theorems to the new result, and (d) by a mixture of these methods. The essence is this—the learning of the situation is not completed with the initial obtaining of a solution. The pattern of the solution must be organized.

The organization, according to conditioning, is organized through drill of many like situations. Actions that lead to success and satisfaction tend to recur. The more they recur the clearer the whole action. As seen above, Dewey insists on a logical deduction. Field psychology says that it is a meaningful analysis of the relations that will give the desired organization. Most mathematics teachers insist that all three procedures are necessary.

The student who obtained his solution to the quadratic equation would probably first verify his answers, then he would proceed to generalize his solution, then verify his generalization. His first task is to make some organization such as: The quadratic equation is related to the quadratic trinomial, equating the equation to zero I factor the one member, the product of two factors is zero only if at least one factor is zero, equating each factor in turn to zero I have transformed the quadratic into two linear equations, I know how to solve these, the two answers satisfy, a quadratic equation has two roots. It is only through such thoughtful organization that he can avoid such errors as equating each factor of $(x - 2)(x - 3) = 6$ to the number 6. This concept is entirely outside of the logical organization obtained from the original solution.

An organization sometimes reveals that the solution to the problem was a fortunate accident, and not a general pattern for solution. Suppose a student is confronted with a number of quadratic equations of the form $x^2 - a^2 = 0$. He writes $x^2 = a^2$. He takes square roots of both sides being very careful to use both positive and negative roots. Then he generalizes: Keep x^2 on one side, get the other terms on the other side, take the square roots of the latter side. This works until he is confronted with $x^2 = 4x + 16$. Then he is lost. The generalization did not really solve his problem. This suggests that for the most part we set up our learning situations to avoid particular solutions right from the start. The problem should lead to a generalization that has the greatest number of possible applications if it is to be of value in new learnings.

The fifth step of reflective thinking is verification, precising, and observation. If the learning has been accomplished, it is ready to be used in new experiences. C. I. Lewis (10) expressed this nicely when he said, "Knowing begins and ends in experience, but it does not end in the experience in which it begins." It is this application of what we have learned to new experiences that leads to creative learning. This is an important aim of mathematics teaching, but an aspect of learning about which most psychologists have little to offer. But if we have learned how to learn through the steps one to four, it would appear that this fifth step is to a large extent the reapplication of these four steps in a new problem with special attention to the use of the material learned.

The person who has learned and organized the general solution of a quadratic equation can now extend it in the new situation $x^4 + bx^2 + c = 0$, or more generally, $x^{2n} + bx^n + c = 0$. This is

THEORIES OF LEARNING

an entirely different type of experience from that in which the original learning took place.

These five steps suggest a modification of the learning diagram of Dashiell as shown below. The steps are illustrated by Roman numerals. The diagram aids in seeing the relationship between the psychological and philosophical explanations of learning. The reader can refer this figure back to the preceding discussion.

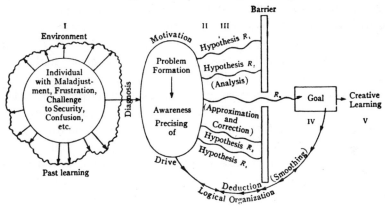

CREATIVE LEARNING

It is essential to note that the logical organization or structuring of knowledge is the final step (never the first step) in developing a basis for future learning. It is also necessary to note that the steps in Dewey's complete act of thought do not necessarily occur in the order one, two, three and, four; nor are they necessarily complete learning. Dewey's explanation of thinking is a final structured explanation of what goes on in learning. A complete analysis must not be made before the learner is aware of the problem. Perhaps after he has an idea of his problem he finds further diagnosis helps further to clarify the problem, and when hypotheses fail, one after the other, he may return and reshape the problem. Further, the process of making a logical structure of the solution may result in the discovery of new and more fruitful hypotheses that give a better structure to the problem. There is a weaving back and forth across the whole learning situation with progressive clarification and elimination of unnecessary details until we finally arrive at a mature, sufficiently well ordered solution. Surely Dewey must have gone through some such procedure in arriving at his final structure of the learning process.

Regardless of how problems of world affairs appeared years ago, today to most people they seem more complex and more serious than ever. There are at least two widely separated theories on how we should behave. The one is to hold or to return to accepted and proved behaviors those that have stood the test of time. In this case our problem is not to unravel a new pattern of living, but to resolve new conditions to old patterns. The other theory is that a changing culture (new conditions) demands a new pattern of living. We are now in one of the ever emerging stages of maladjustment. We must therefore diagnose and clarify our problem, and make hypotheses. We must experiment, generalize, and deduce solutions; we must verify and "precise" our learning. We must be inventive and creative, whether it be in mathematics, social studies, government, physics, or art. We must seek solutions and take responsibility for whatever action our solutions initiate. This is the scientific attitude. This is the way to new knowledge. The key is learning how to learn.

Jacques Hadamard (5) in his *Psychology of Invention in the Mathematical Field* attempted to analyze creative thinking. As he saw it, it comprised four stages.

The first three steps of Dewey's analysis may be taken as the first or the *preparation* stage of creative thinking (step five). To create we must first solve many problems within the field. We must learn facts, skills, attitudes, habits, relationships, and be thoroughly versed in the mathematics. Then a new and unsolved problem may arise in the field. For example, a student may be asked to create a method of drawing a graph of a quadratic function for complex values of the argument which make the function real. It has never been solved so far as the learner is aware. He brings to bear all the past learning and procedures of learning, yet fruitful hypotheses are not forthcoming. To aid the student in this is the function of our teaching.

It is then suggested that for the time being the problem be allowed to rest from conscious attack. This can be looked upon as the latter part of Dewey's step three, and is called the *incubation* stage. Here the subconscious or unconscious, of which we know little, except that it plays an important part in our behavior and probably operates in the inner part of the brain, takes over. Our ideas (hunches), our past knowledge, our concept of the problem, our many unsuccessful hypotheses, and all our knowledge related to the problem, are in, or have had an effect on, the brain and the nervous system. What goes on we cannot explain except to say there is a reorganization, a restructuring of all these elements. The uncertainty and vagueness of action in the unconscious may be one reason why psychologists have, for the most part, omitted it in their theories on learning. Yet certainly something goes on in the unconscious. How do we know it?

By introspection, many, many creators in all fields of knowledge agree that there comes a certain period, frequently unannounced, maybe as we are relaxed before we go to sleep, or as we doze off in a chair, or on the bus or subway, when apparently without effort, insight or emergence of the solution suddenly occurs. This is called the *illumination* stage, when there rises from the subconscious to the conscious all the elements of the problem in an ordered total configuration. It is the beginning of Dewey's step four.

Having achieved insight, the discoverer immediately precises or sharpens his solution. He develops it into a neatly organized, logical, deduced pattern, placing it properly in the larger organization of mathematical knowledge. It is verified and tested as a new learning, as a new contribution to knowledge. This is called the *verification* stage. It is analogous to Dewey's steps four and five. Thus, *preparation, incubation, illumination,* and *verification* as stages in creative thinking have their counterpart in both the philosophic and psychological explanations of learning.

There is nothing new in this process of creative thinking that does not occur in any problematic approach to learning. All such learning is creative. There are successive and overlapping steps which occur in all fields of human endeavor that are making new contributions to knowledge. Insofar as learning how to learn, or problem-solving procedures are general, we can make a contribution in the mathematical field by stressing procedures of learning as much as we now stress the outcomes, without, however, detracting from the securing of necessary skills and facts. Perhaps our lack of creative learning may be traced to our lack of teaching so as to have students learn by problem-solving.

THE ELEMENTS IN LEARNING—SUMMARY

How we learn is described in part by physical processes, by psychological aspects of behavior, and by philosophical considerations. In all descriptions there are present the elements of maladjustment, insecurity, dissatisfaction, motivation, drive, set, emotional disturbance, diagnosis, problem realization, preparation, recall, associations, trial and error (approximation and correction), analysis, hypothesis formation, incubation, solution, insight, goal attainment, illumination, structure formation, smoothing of goal route, precising, deduction, logical organization, and verification. This chapter shows in some manner how these terms are related and what they indicate for classroom practice.

The questions raised by this discussion are numerous. How is the organism motivated? How does it form its attitudes and habits? How are the senses related to learning? What is a concept and how are they formed? To what extent does language aid the learning of mathematics? Is learning achieved at one trial, or is practice a necessary requirement of learning? Does the procedure of learning transfer to later learning and learning in other fields? Just what is a problem, and how does the organism learn to solve problems? Do learners differ in ability to achieve, and if so, how do we provide for various rates of learning within a given class? Must learning be planned or is it a hit or miss affair? If answers to these questions are available, how can a teacher make sensible use of them in his daily instruction?

The rest of this book seeks some answers to these questions.

Bibliography

1. COLE, LAWRENCE E., and BRUCE, WILLIAM F. *Educational Psychology*. Yonkers, N. Y.: World Book Co., 1950.
2. DASHIELL, JOHN. *Fundamentals of General Psychology*. Third edition. Boston: Houghton-Mifflin Co., 1949.
3. DEWEY, JOHN. *How We Think*. Boston: D. C. Heath and Co., 1910.
4. GATES, ARTHUR I., and others. *Educational Psychology*. Third edition. New York: Macmillan Co., 1949.

5. HADAMARD, J. *Psychology of Invention in the Mathematical Field.* Princeton, N. J.: Princeton University Press, 1949.
6. HARLOW, HARRY F. "The Formation of Learning Sets." *Psychological Review* 51: 65–66; January 1949.
7. HILGARD, E. R. *Theories of Learning.* New York: Appleton-Century-Crofts, 1948.
8. JUDD, CHARLES H. *The Psychology of High School Subjects.* Boston: Ginn and Co., 1915.
9. KÖHLER, WOLFGANG. *The Mentality of Apes.* Harcourt Brace and Co., 1927.
10. LEWIS, C. I. "Experience and Meanings." *The Philosophical Review* 43: 134; 1934.
11. MCGEOCH, JOHN. *The Psychology of Human Learning.* New York: Longmans, Green and Co., 1942.
12. MURPHY, GARDNER. *Personality, A Biosocial Approach to Origins and Structure.* New York: Harper and Brothers, 1947.
13. NATIONAL SOCIETY FOR THE STUDY OF EDUCATION. *The Psychology of Learning.* Forty-First Yearbook, Part II. Chicago: Distributed by the University of Chicago Press, 1942.
14. NATIONAL SOCIETY FOR THE STUDY OF EDUCATION. *Learning and Instruction.* Forty-Ninth Yearbook, Part I. Chicago: Distributed by the University of Chicago Press, 1950.
15. SHERIF, MUZAFER. *An Outline of Social Psychology.* New York: Harper and Brothers, 1948.
16. THORNDIKE, E. L. *Human Nature and the Social Order.* New York: Macmillan Co., 1940.
17. THORNDIKE, E. L. *New Methods in Teaching Arithmetic.* New York: Rand-McNally, 1921.
18. THORNDIKE, E. L. *Principles of Teaching Based on Psychology.* Mason-Henry Press, 1906.
19. THORNDIKE, E. L. *Psychology of Arithmetic.* New York: Macmillan Co., 1922. p. 140.
20. UNIVERSITY OF MINNESOTA. *Learning Theory in School Situations.* Studies in Education No. 2. Minneapolis: University of Minnesota, 1949.
21. WERTHEIMER, MAX. *Productive Thinking.* New York: Harper and Brothers, 1945.
22. WOODRUFF, A. D. *The Psychology of Teaching.* Third edition. New York: Oxford University Press, 1951.
23. YOUNG, J. Z. *Doubt and Certainty in Science.* New York: Oxford University Press, 1951.

2. Motivation for Education in Mathematics

Maurice L. Hartung

WHAT impels a person to learn? What starts him into the learning process in a particular situation, and how is his learning activity regulated? These and similar problems are commonly suggested by the term "motivation." They are among the central problems in a theory of learning. To guide learning efficiently, the teacher must have practical answers to such questions.

Psychologists have had much to say about motivation. The problem is a complex one, and psychological theory lacks clarity and precision of the kind the mathematician likes. Many of the terms used, such as "drive" and "set," seem to be loosely defined. Most teachers of mathematics find these discussions hard to understand. Fortunately, in this chapter it will not be necessary to consider in detail the origin of motives, the mechanism of motivation, or issues arising from different theories of learning. However, it will not be possible to avoid using some terms from those parts of learning theory which deal with motivation. First, therefore, attention will be focused on the meaning to be given a few key words, including the terms "motive," "goal," and "incentive." Second, the nature of a few motives useful in teaching will be very briefly indicated. The list includes purposes, interests, attitudes, the need to maintain and build self-esteem, the need for affiliation, and the need for approval. These motives will then be discussed at greater length with special attention to their use in mathematical education. The role of meaning and understanding in motivation will next be considered briefly, and the chapter will conclude with a set of criteria for judging motivational methods and devices.

The emphasis here will be upon a critical survey of general types of methods and devices in relation to modern learning theory. No attempt will be made to give detailed descriptions of methods and devices which have been used for motivation. A very extensive set of suggestions and references may be found in the

Eighteenth Yearbook of the National Council (13), the pages of the *Mathematics Teacher*, and other sources in the professional literature of mathematical education (4).

THREE KEY TERMS

Motives. When a student learns, it is assumed that conditions or states exist within him which initiate and regulate his activity. These are called "motives." Other words often used in connection with motivating conditions are the following: drives, sets, wants, needs, interests. For discussions of these terms and their special connotations see 15:293ff; 14:16–17, 38–39. Among the different motives commonly recognized are hunger, thirst, and sexual urges. These are considered basic or primary motives. They are inborn, universal, and connected with survival. Interests, attitudes, and purposes are types of acquired or secondary motives. They are learned and individualized. For example, the particular interests which serve as motives for one person may not activate another (15:297–98).

Students and teachers may both have motives of which they are not conscious. Also, a person's motives may conflict with one another. His behavior then tends to be determined by whatever motive is dominant. Secondary motives through education may become stronger, at least temporarily, than the basic motives. Thus an individual may continue working on something which interests him even when he is hungry and food is available. Teachers are more concerned with the development and strengthening of secondary motives than with primary motives. Also, teachers must rely mainly upon secondary motives in guiding the learning of students.

Goals. Most descriptions of the learning process make use of the concept of goals. In this discussion, the term "goal" will be taken as undefined. One can say that it refers to what the learner is seeking or wants to attain, but this probably does not clarify the notion for most people.

The responses of the learner in the learning situation are selective. This means that the student responds to some aspects of the situation but not to others. Some responses that are made are not tried again, and other responses that are relevant to the learner's goal are selected. It may be said that "the behavior is directed toward the attainment of the goal." Thus the concept of a goal, when related to motives, may be used to explain how motivated behavior is given direction.

Incentives. Discussions of motivation sometimes use the term "incentive." This term refers to an object or condition which is external to the learner, but which will satisfy a motive which is operating. When, for example, hunger is the motive, food may be an incentive. If the need for recognition and approval is motivating a student, a high examination score may in some cases be an incentive. A high score would probably win approval from his parents, teacher, and certain other students. However, at the same time it may arouse jealousy or resentment, and thus set him apart from a particular person or group whose approval he desperately wants at the moment. The goal is recognition and approval, and a high score may or may not satisfy the motive and be an incentive. Teachers commonly use incentives in their efforts to contact and make use of the motives of students. Teachers hope that incentives will help students move in the direction of their goals.

It should be noted that motives are often stated in terms of goals or incentives. Suppose, for example, the goal is to possess the approval of others. Then the motive may be stated as "the need for approval." Again, a dictionary may say that hunger is "the painful sensation or state of exhaustion caused by need of food." In this case the goal is release from painful sensation, which is internal. The motive, of which hunger is really only a symptom, is conveniently stated in terms of the incentive as "the need for food."

MOTIVES USEFUL IN TEACHING

A number of different motives have been mentioned above. Also, attention was called to the fact that some motives are more suitable than others for practical use in teaching. The most important of the types of motives with which teachers are concerned may now be discussed briefly.

Purposes. Purpose may be defined as the intention to seek a relatively specific goal. For example, when a boy decides he is going to become a mechanic, a doctor, or a mathematician, he may be said to have arrived at a well-defined purpose as to his career. Thereafter, much of his behavior may be directed toward achieving his goal, and so his purpose acts as a motive.

Learning is often greatly influenced by the student's awareness of his need for what is to be learned, or of its potential usefulness and value to him. One of the major responsibilities of the teacher is to help students formulate well-defined goals, and to encourage students to resolve to reach these goals—in short, to help them adopt appropriate purposes.

Interests. The term "interest," which occurs very frequently in educational literature, is used in connection with behavior such as the following: the person wants more of something—he seeks it voluntarily and repeatedly, he stays with it for a period of time; he expresses a preference for or states that he enjoys the activity, and he may recommend it to others. This sort of behavior suggests that a condition exists that satisfies the general definition of a motive. At any rate, teachers are concerned with interests in connection with the problem of motivation.

Interests are always directed toward something—a student is said to be interested in sports, in mathematics, in music, or in a hobby like performing magic. Teachers are frequently advised to investigate the individual interests of their students and to capitalize upon these interests in guiding instruction. Schools also have the responsibility of developing new interests and deepening old ones.

Attitudes. The term "attitude" is commonly used to refer to an idea or set of ideas which have emotional content. Attitudes, like interests, are always directed *toward* or centered *about* something. Thus a person may have attitudes about war, or people of other races, or a school subject like mathematics, or a particular teacher. These ideas are not the same as cold facts about which he does not care. Rather, they are beliefs about the object or subject, or prejudices favorable or unfavorable to it, that really matter to him personally.

Attitudes influence behavior and hence act as motives. They are learned and, in turn, they often make new learning easier or harder to acquire. One of the chief obstacles to the effective learning of mathematics is the unfavorable attitude toward the subject which has been acquired by many students.

Need to maintain or build self-esteem. Much of a person's behavior is guided by his feeling of personal worth. He tries to maintain and build his self-respect, and he tends to avoid behavior which lowers it. The things that contribute to self-respect differ greatly from person to person. Behavior by a student which lowers the *teacher's* respect for him may actually contribute to or build up his *self*-esteem (6:84). Thus a pupil who found long division confusing and difficult decided never to work on another long division example, and informed the teacher of his decision. This pupil tried in this way to avoid a situation in which he was experiencing repeated failure with accompanying loss of self-esteem. The use of sarcasm and ridicule by teachers is frowned on by supervisors and other experts in methodology because, among other reasons, it tends to lower rather than build up students' self-esteem.

Need for affiliation. Most people gain some satisfaction from playing a constructive role in a functioning group. The sense of *belonging* to a family, a gang or crowd of his peers, or to other groups, is important to a person. The school provides a group or groups in which the student is, physically, a member. If, however, he is not really accepted as a person by other members of the group, his need for affiliation may lead to behavior which interferes with the learning task set by the teacher, and with the learning of other pupils. Often the attention-getting activity of students is designed to gain status with the group. Hence even if teachers do not exploit this motive constructively, they must deal with the behavior which is associated with it (14:41).

Need for approval. People usually put value on receiving the respect and approval of others. They want to be considered important—to be recognized. This need for approval is obviously related to the need for self-respect and the need for affiliation. It can become a very strong motive for learning, and it is widely used by parents and teachers in promoting learning.

Interrelationships of these motives. The motives discussed above are not to be considered as sharply distinct or independent. Several of them may be operative in the same individual at the same time. If so, they may tend to assist each other, or as noted earlier, they may conflict. When, in guiding learning, conflicts can be reduced and several motives can be induced to operate together to produce a desirable type of behavior, it is clearly advantageous to do so.

Discussions of teaching frequently stress the importance of motivating the student. Some writers seem to believe that it is the job of the teacher, or of the textbook, or ideally of both acting jointly, to motivate the learner. The point of view implicit in these statements is that motives are to be *put into the student* prior to or in the early phases of a learning experience. This view is not in accord with the definition of motives given above. Rather, the definition suggests that the motives the student *already has* are to be contacted, or perhaps aroused at the outset of a learning experience. That is, the teacher utilizes existing motives in promoting a particular learning experience. Most learning experiences, however, involve multiple outcomes. Thus while he is learning mathematical facts and skills the student is also acquiring attitudes, interests, and purposes in relation to the subject. Later, these may act as motives in relation to new learning experiences. These motives may be either favorable or unfavorable to the development of the behavior desired by the teacher.

PURPOSES AND GOALS

Modern methods of teaching recognize the value of giving attention to the students' purposes. This involves the identification and clarification of appropriate goals by both teacher and student. Purposes growing out of the life goals of students, and particularly those related to vocational and homemaking goals, are effective with students in secondary schools and colleges. Goals of this sort are remote and relatively ineffective with elementary school pupils. Their goals tend to be of a sort more immediately attainable. Teachers usually find that purposes growing out of the activities of daily life are effective with these younger pupils. At all levels, however, the extent to which the goals are realistic and the possibility of successful achievement must be considered.

Contacting life goals. In mathematical education, methods and materials commonly used to connect learning experiences to possible life goals of the students include those in the following list derived from a survey of the literature (20:52).

1. Assigning problems which are real to the students here and now, such as handling the mathematical computations of the school cafeteria (16), and studying their own social habits statistically (3). This type of procedure was sometimes referred to as *functional* teaching—teaching only those things of which the students could make immediate use.

2. Pointing out that the methods of attacking problems in life are much the same as the methods of attacking problems in mathematics.

3. Showing how mathematics helps us to understand our environment.

4. Making practical applications of the mathematics being studied and by pointing out the extent of mathematical applications in domestic and vocational life.

5. Having students conduct surveys of their own to determine how much mathematics is used in daily life. For example, one writer described how pupils conducted a survey of popular magazines to determine the extent to which mathematics and mathematical terms were used in them (22).

6. Emphasizing the importance of mathematical thought in philosophy and as a factor in the evolution of civilization.

7. Using field trips for mathematical study of the environment.

8. Using visual aids (posters, photographs, slides, and other materials) which bring out the importance of mathematics in practical applications (10).

The major emphasis in this list is upon mathematics as it is used by and has bearing upon the citizen in his ordinary life activities. If the student accepts the idea that he needs to know mathematics for successful living, he recognizes a goal and may formulate a purpose—namely, to learn more mathematics. It will not be sufficient, however, to give him a general discussion on his need for mathematics—a sort of pep talk— at the beginning of the year and occasionally thereafter. To be fully effective in motivation, this general purpose must be related constantly to the specific topics or skills which form the content of the daily learning experiences. Moreover, these topics and skills must be genuinely useful to most citizens. It is rarely possible to maintain for any length of time the illusion that a topic or skill is useful when it in fact is likely to be of little or no use. Awareness of the usefulness of mathematics in daily life is rarely effective in motivation unless the material being studied is clearly relevant to this goal.

Mathematics is used extensively by many types of trained workers and in numerous technical professions. This suggests a second type of emphasis useful in attempting to relate mathematics to the goals of students. Information as to the kinds and amounts of mathematical knowledge and skills actually needed can be made readily accessible to them, and they can be encouraged to study it. Under the guidance of a sympathetic teacher, such study may help to clarify goals and purposes.

Both the above emphases have been extensively discussed in professional books and magazines. They were also ably developed in the *Guidance Pamphlet in Mathematics for High School Students* (9). This document is particularly valuable for several reasons. In the first place, it is addressed to the high-school student directly, and it aims to assist him in thinking about his own purposes rather than to state purposes which he is expected to accept uncritically. In the second place, it avoids the common error of overstating the case. Many teachers in their personal enthusiasm for mathematics have "scared" students away from certain courses or types of study by overstating the amount of mathematical training or degree of skill demanded in a later course or a profession.

A relatively small but extremely important group of students will ultimately become professional mathematicians. Motivation is not an urgent problem with this group, but they also can profit from guidance which enables them to learn of the ever-growing scope of the opportunities in the field. Recently there has been a rapid increase in guidance materials designed for use in colleges (18).

Much of the literature on the importance of mathematics in college work and the professions is based on the opinion of those already in the profession or of mathematicians. The value of such views is not questioned, but they may be less effective with high-school students than the views of their contemporaries would be. An indication of what these views may be was obtained in 1943 by Crawford by means of a follow-up study of 424 high-school graduates of 1942 (7:47). These students regarded English as the most helpful of all subjects, with mathematics running a close second Moreover, 43 per cent (including twice as many boys as girls) indicated they would take more mathematics if they could replan their high-school programs in view of their present situations. Data like these from recent graduates are especially useful in helping students clarify their ideas about the potential role of mathematics in their own lives.

Realism in selecting goals. When a goal or incentive has been attained, and a motive or motives thereby satisfied, we commonly say that the learning experience was successful. When a goal or incentive has not been attained, in spite of effort to do so, we consider the learning experience a failure, or at least a partial failure. Now if a goal has been properly chosen, certainly what both teacher and student want is success in achieving it. It is, therefore, not necessary to espouse any particular psychological theory as to the role of success in effecting learning in order to take a position in favor of so arranging conditions that success is possible and probable.

The goals selected must be realistic in the sense that they are attainable. If the teacher attempts to set goals which the students cannot attain, or if the students themselves are too optimistic in choosing goals, excessive frustration and disillusionment usually result. On the other hand, if the goals are too easily accessible, there is insufficient challenge and less than maximum achievement. Under ideal conditions teacher and students working together clarify the purposes of the learning experience on which they are embarking, and they attempt to formulate a reasonable set of goals for the class as a whole and for individuals. Moreover, as work proceeds they re-examine the purposes and adjust them, if necessary, as the situation unfolds. The learning situation thus established is quite different from that in which the teacher makes a fixed, predetermined assignment. When the student is given an opportunity to participate with a group in formulating purposes, a contribution may be made toward satisfying his need for affiliation. In this way the general learning situation is improved. By participating in the goal-setting process, the student is not only more likely to be fully aware of what the goal is, but also more ready to adopt it as his own, so that it can act as a purpose or motive.

INTERESTS AND INTEREST-AROUSING DEVICES

Interests are motives that almost all teachers try to use to promote learning. The literature on mathematical education contains numerous references to the importance of arousing the interest of the students, and dozens of suggestions as to means of doing so. Unfortunately, many of these suggestions seem to stem from a relatively narrow conception of the nature of interest in mathematics.

Genuine interest in mathematics probably depends basically upon the problem-solving aspect of the subject. Problems, once recognized or sensed, leave an individual in a state of perplexity, uneasiness, or tension until they are solved. When a solution has been found, tension-reduction and satisfaction results. If mathematics is properly taught, it presents the student with an abundance of problems, and it also provides him with certain general modes of thought and a supply of techniques which enable him to attack these problems successfully. With each successful solution he receives a dividend of satisfaction—he feels good when he gets the answer. As a result, he seeks more experiences of the same kind, and displays other desirable types of behavior which were described earlier in defining interest.

As the student grows in mathematical maturity, he obtains satisfaction also from comtemplation of the power of his methods and the sharpness and the beauty of his tools. The term "appreciation" is often used in this connection. The behavior is relevant to interest, however, because it leads the student to seek more experiences with mathematics, to discuss it favorably with other people, and to value it for what it does for him personally.

A good many of the devices recommended for arousing interest seem to be based on the assumption that mathematics itself is uninteresting and hence learning must be encouraged by extra- .neous elements such as the mystery of a puzzle or the competition of a game. When, however, the subject preferences of children are investigated by objective methods, mathematics is often found to rank high in the list. In a recent study of the preferences of thousands of fifth-grade children in New England, arithmetic was ranked by the girls second and only slightly below reading (5). The boys gave first place to arithmetic. When the rank order of *disliked* subjects was examined, arithmetic was placed second and only slightly under geography by the girls, and was ranked fourth in disliked subjects by the boys. Arithmetic ranked first as a favorite subject of the 543 teachers of these fifth-grade children. The children of a given teacher tended to follow her preference. There is no evidence that these 543 teachers were using many special devices to arouse interest in arithmetic. This tendency for mathematics to be either very well-liked or heartily disliked has been found in other investigations also.

In another similar investigation 2164 girls and boys in three school systems—one urban, one rural, and one mountain—placed arithmetic first among their preferences in Grades IV through VIII. In high school, however, mathematics was ranked third, with English and Social Studies above it (12:34).

When some of the devices often recommended for use to create interest are examined critically, their limitations for this purpose become apparent. The use of material on the history of mathematics, for example, is frequently suggested, and many textbooks contain some materials of this kind. What kinds of behavior do these materials elicit? First, the student reads some biographical information about a famous mathematician, or some facts from the history of a particular topic. This activity may or may not give him any satisfaction or stimulate him to want to learn more mathematics. Reading about mathematics or mathematicians is rarely a problem-solving type of activity, and thus lacks some of the motivation of problem situations. Historical materials are probably most effective with students who have already developed considerable interest in mathematics, and they do very little for those whose interest is meager or non-existent. It is true that stories of some of the dramatic episodes in the history of mathematics may stimulate interest on the part of students for whom the subject has little appeal. They should certainly be used whenever possible. The difficulty is that they cannot be used often enough to provide a cumulative effect and build a sustained interest in mathematics itself.

Other devices commonly suggested include mathematical tricks, games, puzzles, and other recreations. All of these have at least a temporary appeal for many students. To secure a solution calls for behavior which has some elements in common with problem-solving in general—there is perplexity and challenge at the outset, and satisfaction at the successful result. However, tricks generally demand only the giving or following of a set of directions, and puzzles are not usually solved by straight-forward methods. If tricks and puzzles are overemphasized, the students are likely to get a distorted idea of the nature of mathematics. They think that mathematics consists largely of a bunch of tricks. Some teachers have such faith in the motivating power of tricks and puzzles that they use these terms in discussing even the standard processes of mathematical work. Thus they may refer to the reduction of fractions, or the replacement of $x^2 + 2xy + y^2$ by $(x + y)^2$, as a "trick." The motivating power of such devices and the wisdom of using them are both open to question.

If a theoretical explanation of a trick or puzzle is accessible to the student, the discovery of it may of course become a genuine problem. Investigating whether a trick which works for a particular set of numbers will work for others, or in general, may be much more interesting than many practical problems. It is to be noted, however, that in this case the interest is derived not from the trick as a trick, but from the more general mathematical or problem-solving behavior that is evoked.

Most of the games used in mathematics courses are adaptations of other non-mathematical games and are designed to practice skills rather than provide problem-solving experience. Their effectiveness in motivation depends upon factors such as the extent to which playing them with others satisfies the need for affiliation, or winning the game satisfies the need for recognition. The mathematical aspects are frequently a minor facet. Similarly, the writing and staging of a mathematical play may stimulate a considerable amount of valuable learning activity, but the interest is often centered more upon the dramatic activity with classmates than it is upon mathematics.

In this connection a few comments on mathematics clubs may be appropriate. Although club programs often rely heavily upon recreational activities in mathematics, the danger that the members will acquire a distorted idea of mathematics may be discounted. This conclusion is based on the fact that membership in the club is almost invariably voluntary. Only students who already have rather well-developed interests participate. These students have had many successful and satisfying experiences in regular mathematics classes. Their ideas about the nature of mathematics are, on the average, more rounded and mature than those of their regular classmates. Club programs can further develop the interest and understanding of these students by offering experiences suited to their special talents. Motivation is not much of a problem for the club sponsor, but it is of crucial concern in the regular classroom. The teacher of mathematics may wish that all students could be as well motivated as the club members are. At the same time, the experienced teacher knows that materials appropriate for a mathematics club may be ineffective in ordinary classes.

Many teachers think of films, filmstrips, models, bulletin-board displays of clippings, pictures, posters, and the like, primarily as motivating devices. These should, however, be regarded primarily as learning aids, and their role is more like that of the textbook. That is, they serve to clarify concepts and processes, explain how a principle works, or bring background situations and data from the outside world into the classroom. Because they help students get increased meaning and understanding, these learning aids tend to increase interest and promote the development of favorable attitudes. Moreover, they bring greater variety of experience into the learning situation. Their concrete and visual characteristics attract attention, and it must be remembered that attention precedes interest. Motivation is almost invariably improved when they are used. If these materials were primarily designed for motivation, a longer discussion of them would be appropriate here. The greater importance of these materials as learning aids justifies the inclusion of a much more extended discussion of them in a later chapter.

The teacher of mathematics who wants to help students acquire a deeper interest in the subject may legitimately make use of all of the devices mentioned above. Variety will be more effective than the excessive use of one or a few devices. This is true not only because the over-use of a single device of this kind dulls its appeal, but also because some students who are not interested by one device may be reached by another. In the last analysis, however, the teacher must remember that these are, after all, only devices. They are, in one sense, a diversion from the main stream of mathematical learning. The development of a deep and permanent interest in the field is more likely to be fostered by regular and successful experience in solving representative mathematical problems.

ATTITUDES AND SUCCESSFUL ACHIEVEMENT

All teachers are familiar with the fact that people usually have a definite set of attitudes toward mathematics. These attitudes may be quite favorable or strongly unfavorable, but they are rarely neutral. Parents often say: "I am not surprised that my child isn't doing well in mathematics. I had a terrible time with it myself while I was in school. I never liked the subject." Since attitudes are often taken over from others, children are likely to acquire the attitudes of their parents. They may also be influenced by awareness of the attitudes of their schoolmates and teachers.

In a recent study by Dutton, written statements of attitudes toward arithmetic were collected from 211 prospective elementary-school teachers (8). Only 26 per cent of the statements were favorable to arithmetic, and 74 per cent were unfavorable. The language used was expressive and emotional, revealing deep-seated attitudes that had persisted from childhood. Prominent among the causes given for the unfavorable attitudes were lack of understanding, failure to provide enough application to life and social usage, poor teaching techniques, poor motivation, and feelings of inferiority and insecurity. Some of the statements clearly revealed that these students had been influenced by the attitudes of their parents.

Successful Experiences. Students often acquire attitudes as a result of repeated experiences of a similar type. In particular, it is well known that repeated successful experiences with mathematics may lead to favorable attitudes, and similarly, repeated experience which is unsuccessful and unsatisfying is likely to lead to the development of an unfavorable attitude toward the subject. For example, in a study of the attitudes of students in a commercial arithmetic course, Billig found a definite positive relationship between attitudes, as expressed in written statements, and achievement (2).

The attitudes toward school, toward achievement, and toward an education, of 200 children selected to represent certain achievement ability patterns were studied by Kurtz and Swenson (11). They found these attitudes to be more closely related to the students' achievement scores than to their ability (intelligence-test) scores. The same study showed considerable agreement in the attitudes of parents, teachers, and children.

Success in achieving purposes is generally preferred, by students and teachers alike, over failure. Failure is seldom deliberately selected as a goal by normal individuals, and fear of failure is generally not regarded as a desirable type of motivation. Nevertheless failures, in one sense or another, do occur, and some attention must be given to this problem. The argument is sometimes put forward that failures in school are justified as preparation for inevitable failures in life outside school—in other words, the student must learn how to adjust to failure. It should be noted, however, that life outside school gives ample experience in failure, and that it is not necessary for the schools to provide additional experiences. Moreover, in life outside school the individual has some freedom of choice and action and can substitute another and more attainable goal, while schools and teachers tend to force children into repeated experiences of failure to reach the same goal (14:52).

Successful experience helps the student maintain his integrity and self-esteem, while repeated failure tends to tear him down. Successful experience contributes to the building up of attitudes favorable to the task, and of interests in it, while failure contributes to unfavorable attitudes toward the experience and inhibits the development of interest in it. Successful experiences win approval and recognition, while failures often call forth disapproval. Hence if these varied types of motives are to assist purposes and continue to operate in promoting learning, the experiences must on the whole be successful.

Security, order, and system. If achievement in mathematics is reasonably successful, the student can develop certain feelings of security which tend to promote favorable attitudes (1). Children can learn to check their own answers, so they feel confident the result is correct. Few if any other types of content have this characteristic to the same extent as mathematics. It is important, however, that the teacher help the student to see the importance of checking his work in terms of his own satisfaction. Far too often he regards checking as useless drudgery, and such an opinion may lead to an unfavorable attitude. For example, teachers sometimes require pupils to check multiplication examples by long division, and conversely. The checking process then becomes laborious and loses some of its value in contributing to security. In checking such examples most adults prefer to run the risk of repeating an error by going over the work, but they nevertheless satisfy themselves that it is correct. The possibilities mathematics can provide for developing favorable attitudes toward correctness and precision provide one of the strongest arguments for giving it a prominent place in the curriculum. To encourage achievement of this favorable result, teachers must provide the student with appropriate methods of checking and the time to use them. Over-long assignments, which drive the student to cover the material in a hasty or superficial manner, must be avoided. Speed must not be sought at the expense of accuracy.

An unusually interesting study of the attitudes of young children toward mathematics has been reported by Plank (17). The case study method was used with 20 children, some of whom were retarded while others were accelerated in arithmetic. Of special significance is the evidence of the relation of personality characteristics, such as insecurity, anxiety, and rigidity, to performance in arithmetic situations. Among the conclusions occurs the following: "The insecure children show a definite discrepancy between their scores in reasoning and computation in their achievement tests.

They can neither stand the competitive atmosphere that goes with computation nor the emphasis on speed while they are trying to be accurate" (17:263).

• In many non-mathematical situations, the individual may escape with a hasty or partial answer. Often a personalized type of response is entirely satisfactory—for example, an opinion, a statement of like or of dislike for a poem or a musical selection. The student may also be encouraged to produce art products which express his own personality. In mathematics such avenues of adjustive behavior are greatly restricted. Tension-reduction and satisfaction are not so easy to attain. Thus belief in the desirability of precision and correctness must be slowly and carefully developed. If this is not done, premature demands for high precision and correctness may lead to unfavorable attitudes toward mathematics.

The order and system of mathematics, when it is learned with meaning and understanding, usually is a source of satisfaction and security to students. The manner in which a new topic fits into and builds upon previous learning can be made evident. There is increasing complexity of concepts and skills as one unit or course builds upon another, but the number of really basic principles is surprisingly small. At the same time, this characteristic may be a source of extreme insecurity, since if one or more of the basic concepts is not learned at the proper time, a very serious handicap is imposed. In a less orderly and systematic subject, such an omission is less likely to lead to learning difficulties and frustration later. Thus the systematic nature of mathematics can influence attitudes both favorable and unfavorable to the subject.

An appropriate curriculum. The relation of successful experiences to attitudes, and hence to motivation and learning, provides one of the strongest arguments for curriculum reorganizations. It suggests that content too difficult for a particular group of students at a particular time should be relocated at a more favorable time, or in some cases omitted entirely. It demands the use of methods and materials which facilitate learning. It lies back of efforts to care for individual differences by grouping pupils in various ways, by offering two-track programs, by use of differentiated assignments, and by other methods. There is a tendency for the curriculum to remain static while teachers frantically search for devices to make it palatable. Although some of these motivating devices may in appropriate circumstances increase the likelihood of successful learning, they are probably far less effective, in the long run, than efforts to adjust the content of the curriculum to the learner's purposes, interests, and capacities. Most of the chapters that follow are relevant to motivation insofar as they include discussion of ways of accomplishing these ends.

SELF-ESTEEM, AFFILIATION, AND APPROVAL

The need to maintain and build self-esteem, the need for a sense of belonging to a group, and the need for recognition or approval, have all been discussed above in connection with other motives. The teacher of mathematics may help or hinder the student in his effort to meet these needs. In many cases in a learning situation the role of the mathematics will be less prominent than the role of the particular methods that the teacher employs in handling students. In other words, it is probable that the educational philosophy of the teacher, and the influence of this philosophy on student-teacher relations, is more influential than the subject matter being studied.

Incentives. One method that many teachers rely on to promote learning is the use of incentives. These incentives take many particular forms, but rewards, punishments, and competition are usually recognized as general types. The relationship of incentives to such motives as the needs for self-esteem, affiliation, and approval has been extensively discussed, but the desirability of using certain kinds of incentives in school still remains in the domain of controversial issues.

Incentives are used by parents in the home and by workers with youth in the community. Parents reward their children for good behavior, and punish them for bad behavior. Often the reward is only a word like "Good!" a smile, or a kiss; but sometimes it takes a more tangible form such as candy, money, or a new piece of clothing. Punishments are equally varied, but frequently involve the withholding of anticipated rewards. By means of incentives most parents regulate the behavior of their children not only in the home, but to some extent elsewhere including the schoolroom. Similarly, organizations in the community, such as the Boy Scouts and the local recreational association, offer incentives to learning and achievement, usually in the form of badges and prizes. The practice is more or less taken for granted.

In view of these circumstances, it is not surprising to find teachers using incentives to promote learning in schools. The smile of the teacher, praise for a job well done, a piece of work displayed upon the bulletin board, a favorable report to the parents, the publication of the student's name in the honor roll, may all on occasion serve as incentives. Students have learned outside school to expect incentives of some kind to be in evidence.

However, in recent years there has been a strong tendency among educational leaders to challenge many of the common practices of teachers in connection with rewards and punishments. The following quotation from Hilgard and Russell (14) states some of the questions which concern thoughtful educators.

That rewards influence learning is beyond doubt. The question becomes how they can be used appropriately. One of the problems which exists for the teacher is that of the byproducts of reward, when reward is viewed as part of a total social situation. A teacher-planned reward extrinsically related to the learning task is a kind of bribe and may lead to the attitude, "What do I get out of this?" That is, an activity is only worth while for the remuneration it brings in praise, attention, or financial gain. Then there is the question of what happens to those who fail to get the reward. If there is only one prize and many contestants, the problems of the losers are to be faced along with those of the winners. Perhaps the winner will be encouraged through the effectiveness of his reward, but what of the others? Is the price in disappointment to them worth what the gain was for the winner? Rewards are almost always competitive: If everyone receives the same recognition or gets the same mark, then the reward value goes out. Even in the stress on group achievement, the group may be set in competition with another group, in order to maintain the reward value of special status (p. 48).

Marks. Now it happens that the variety of types of incentives admissible in schools is much less extensive than the set available to parents. Candy is rarely considered an appropriate incentive in schools. Money rewards are limited to scholarships and a few prizes. Punishments are restricted as to type and severity. Moreover, the incentives available for use by teachers of mathematics are somewhat restricted in comparison to those of certain other fields. Musical and dramatic organizations give public performances, athletic teams play those of other schools, work in the shops may produce useful objects, and the product of a sewing class may be a new dress. On the other hand, mathematics as often taught lends itself to frequent and precise testing. The correctness of an oral or written answer to a mathematical exercise is usually easy to check. Although test scores and school marks are not the exclusive tools of mathematics teachers, marks have often been one of the principal incentives relied upon in mathematics classes.

School marks, if properly determined, can be useful in a limited way as evaluative summaries. In practice, however, marks usually become incentives also. High marks as incentives may satisfy the need for recognition and approval. They help to build up self-esteem and, if they are truly deserved, add to the student's sense of personal worth. They are symbols of success. Conversely, low marks operate to reduce self-esteem and probably increase rather than satisfy the need for recognition and approval. An unfavorable attitude toward the subject is a common result. Since only a few students of any group ordinarily receive the high marks, the total effect on the group can be more unfavorable to future learning than it is favorable. Moreover, emphasis upon marks as incentives sometimes leads to undesirable forms of behavior such as copying homework and cheating on examinations.

Although many schools have abandoned school marks of the traditional kind at the elementary level, the high schools and colleges will doubtless continue to use them in the years just ahead. Teachers should therefore seek to turn them to maximum advantage. They can, for example, emphasize the evaluative rather than the incentive aspect. They can avoid putting too much emphasis on high marks, as such, and be careful about how they give public praise to the students who earn them. A written note on the student's paper, or a word of congratulation in a private conference, will usually provide adequate recognition and approval. Similarly, teachers can avoid publicizing low marks to the class. In private conference with low-ranking students they can at least try to refrain from using words that express disapproval and lower the self-esteem of the student. They can direct attention to positive aspects by focusing upon diagnosis of difficulties and the determination of steps that can be taken to improve the situation.

Competition. The use of competition is rather commonly considered a means of motivation. However, as one writer states: "Competition itself is not a motive.... Competition is itself motivated—it is a response made to certain motivating conditions and exists because there are others who are similarly motivated" (21:618). For some people, competition may serve as a socially approved form of aggression. Competition is possible because the successful competitor usually receives some form of social recognition or approval and a heightened sense of personal worth.

Although most authorities agree that competition often seems to get results, evidence is accumulating that it also can have harmful effects. It tends to produce excessive individualism. It interferes with cooperative efforts—working with and for the group—and thus comes in conflict with one of the citizenship objectives of the school.

It seems probable that most of the undesirable aspects of competition in schools may be side-stepped if no pressure is put upon the students as a group to enter into competition with one another. If, for example, a superior student voluntarily enters a contest or other form of competition, the situation is quite different from one in which *everyone* in a group is put in a competitive type of situation. Moreover, it seems obvious that a contest between reasonably well-matched individuals may be so controlled by intelligent management that the beneficial effects are at least equal to any possible harmful consequences.

It was mentioned earlier that the use of competition as a motivating device is based on the assumption that the mathematics itself is not interesting, and thus the student requires the extraneous type of motivation implicit in competition. In defending the use of competition in schools, it is well to avoid analogies with the business world where incentives in the form of financial profits are quite acceptable.

In summary, incentives in the form of marks or prizes, and the types of competition which tend to be associated with them, are less favorably regarded today than they were formerly. This is true because educators have become aware of the unfavorable effects of these incentives upon many students. Although such methods enhance the self-esteem and help to fill the need for approval in a few students, they lower the self-esteem of many others. Attention is shifting to a search for better ways of meeting these needs.

One of the newer methods emphasizes the cooperative attack upon quantitative problems by whole classes or through committee work within classes. Thus, when a problem calls for the gathering of data in the community, the library, or by correspondence with people of other communities, opportunity is afforded for cooperative group activity. Each member may assume responsibility for a small but different portion of the work. The results may then be pooled and interpreted as a whole. Such methods are designed in part to contribute to the sense of belonging to a group. Recognition and approval are in terms of the individual's contribution to the group.

The use of methods of this sort is relatively infrequent in mathematics classes, but is growing rapidly in some other fields. Group dynamics is now one of the most active of research fields. These studies have already made clear that the ability of a group to accomplish its purposes depends upon many subtle factors influencing the relationship of the members to each other. In particular, the individual's concept of his role in the group is important. This concept is influenced by his need to maintain his personal integrity, his need for affiliation, and his need for approval. Teachers of mathematics who wish to maximize achievement should probably give at least as much attention to these motives as they now do to the cultivation of the more obvious motives such as interests.

UNDERSTANDING AND MOTIVATION

As time goes on, more and more teachers of mathematics recognize the importance of making the subject meaningful to students, and of helping them understand it. Unfortunately, teachers have not always agreed as what constitutes meaning and understanding, or on how these are developed in students. Mathematical education is now actively engaged in exploring methods of achieving these objectives. Later chapters have much to say about it. Here we shall discuss only the relation of meaning and understanding to motivation.

Experimental investigations which have studied the development of meanings and understandings often report that motivation was easier or better in groups which were helped to acquire deeper meaning and better understanding. It is easy to see why this occurs. All too frequently students who *want* to understand their mathematics have not been given adequate help and encouragement. Failure to reach a satisfying level of understanding results in lowered self-esteem. In contrast, successful efforts to help all students gain meaning and understanding tend to maintain and enhance individual and group self-esteem.

When students gain meaning and understanding as part of their learning, there is a better chance that the elements of favorable attitudes will be formed and new interests aroused. A recent study undertook to develop an instrument for the objective measurement of motivation in mathematics. Scores on this instrument were correlated with a number of other measures including an intelligence test score and scores on several well-known aptitude and achievement tests in mathematics. Among these measures were scores on a test of *understandings* in arithmetic. The correlation coefficient obtained in this last case was much larger than the others. When understanding occurs, better motivation tends to go along with it (19). This in turn may affect later learning experiences in ways which increase the level of achievement and also result in total attitudinal and interest patterns of a desirable sort.

Most discussions of the desirability of developing meaning and understanding in connection with mathematical learning emphasize the mathematical aspect. It must be remembered that the meaning of an experience includes more than the mathematical concept or skill which is the supposed object of attention. The purposes, attitudes, and other motives of the learner influence *the meaning of the experience for him*. The mathematical meaning does not change from person to person or from day to day. The total meaning of a mathematical experience does change and is affected markedly with the circumstances of a given situation. Much of the earlier discussion of this chapter has been related to meaning in this larger sense of the term.

In recent years many teachers have had opportunities at conventions to see mathematical films and filmstrips, models, charts and other learning aids exhibited. They have read articles about such materials in professional magazines. In institutes, workshops, and summer school classes, they have had experience in making a few learning aids of their own. The result has been a great upsurge of enthusiasm for these materials, and renewed energy for teaching mathematics in general. In a word, the motives of these teachers were aroused. They acquired new interests, new goals, new incentives, and new purposes. One can hardly believe that this all came about because these teachers saw, or made for themselves, a few teaching aids. This phenomenon can be explained only by recognizing that these teachers acquired new meanings and understandings from their experiences, or new insight into methods of teaching. They also acquired renewed interest as a result of taking a problem, such as the construction of a learning aid, and solving it to their own satisfaction. Moreover, they found ways of satisfying their need for affiliation as they with other teachers worked in groups. Their feeling of personal worth was enhanced through their participation in groups devoted to professional improvement and satisfying social activity. Teaching as a whole had new meaning for them, and it is almost certain that those who have had such experiences are better teachers as a result.

CRITERIA FOR EVALUATING MOTIVATING PROCEDURES

The teacher in preplanning for the work of a class cannot ignore the problems of motivation. Many possible procedures may come to mind. Some of these will be rejected while others will be accepted, at least tentatively, as suitable. Prior sections of this chapter have indicated some of the basic factors to be considered in making such decisions. It may be helpful, however, to list here a set of questions that might be asked in this connection (20).

1. Is the proposed procedure likely to be effective?
    a. Does it draw upon motives actually present in the learner?
    b. Is it designed to utilize a combination of several motives in the learner?

c. Is it appropriate for the age level of the learner?
   d. Is it based upon recognition of a goal by the learner, and does the learner believe he can achieve the goal?
   e. Does it motivate many students or just a few?
   f. How long is the motivation likely to persist?
2. Is the motivation of a desirable type?
   a. Does it lead the student to value the learning experience itself rather than external rewards?
   b. Will it widen and deepen the interests of the learner?
   c. Does it tend to develop desirable attitudes toward the content or skill and toward the teacher?
   d. Are the goals which are set actually attainable?
   e. Does the motivation tend to strengthen attitudes necessary for democratic citizenship?
   f. Is the motivation consistent with the promotion of good social relations between students?
3. Is the procedure practicable?
   a. Is the required expenditure of time and money within the means of the school?
   b. How well can the procedure be controlled in practice?
   c. Does the teacher know how to administer the procedure?

In listing these criteria for judging motivational procedures, there is no intention to suggest that the questions in the list are to be formally answered in connection with every proposed motivating activity. This is obviously impractical. Rather, the list is to be viewed as suggesting the kinds of questions teachers should have in mind as they think over possible procedures which occur to them. These questions may also be useful in evaluating motivational procedures and devices described in the professional literature or in professional meetings.

In conclusion, we may reflect that the interests and attitudes of most children when they enter school are favorable to learning mathematics. Some of them learn to *like* it. Others learn to *dislike* it. To change an unfavorable attitude once formed into a favorable one is a difficult assignment. Efforts on the part of teachers to arrange conditions so that unfavorable attitudes are not learned will, in the long run, probably pay generous dividends.

Bibliography

1. ADLER, ALFRED. *The Education of Children.* New York: Greenberg Publishing Co., 1930. p. 309.
2. BILLIG, ALBERT L. "Student Attitude as a Factor in the Mastery of Commercial Arithmetic." *The Mathematics Teacher* 37: 170–72; April 1944.
3. BOYCE, G. A. "Injecting Life into Mathematics." *Junior-Senior High School Clearing House* 10: 483–85; April 1934.
4. BUTLER, CHARLES H., and WREN, F. LYNWOOD. *The Teaching of Secondary Mathematics.* New York: McGraw-Hill Book Co., 1951. p. 126–57.
5. CHASE, W. LINWOOD. "Subject Preferences of Fifth Grade Children." *Elementary School Journal* 50: 204–11; December 1949.
6. COREY, S. M.; FOSHAY, A. W.; and MACKENSIE, G. H. "Instructional Leadership and the Perceptions of the Individuals Involved." *Bulletin of the National Association of Secondary School Principals* 35: 83–91; November 1951.
7. CRAWFORD, JANE ELIZABETH. "A Survey of High School Graduates of 1942." *School Review* 53: 44–49; January 1945.
8. DUTTON, WILBER H. "Attitudes of Prospective Teachers Toward Arithmetic." *Elementary School Journal* 52: 84–90; October 1951.
9. NATIONAL COUNCIL OF TEACHERS OF MATHEMATICS. *Guidance Pamphlet in Mathematics for High School Students.* Final Report of the Commission on Post-War Plans. November 1947. Washington 6, D. C.: The Council, 1947.
10. JOHNSON, D. A. "Vitalizing Geometry by the Use of Pictures." *School Science and Mathematics* 38: 1032–1034; December 1938.
11. KURTZ, JOHN J., and SWENSON, ESTHER J. "Student, Parent, and Teacher Attitude Toward Student Achievement in School." *School Review* 59: 273–79; May 1951.
12. MOSHER, HOWARD H. "Subject Preferences of Girls and Boys." *School Review* 60: 34–38; January 1952.
13. NATIONAL COUNCIL OF TEACHERS OF MATHEMATICS. *Multi-Sensory Aids in the Teaching of Mathematics.* Eighteenth Yearbook. New York: Bureau of Publications, Teachers College, Columbia University, 1945.
14. NATIONAL SOCIETY FOR THE STUDY OF EDUCATION. *Learning and Instruction.* Forth-Ninth Yearbook, Part I. Chicago: Distributed by the University of Chicago Press, 1950.
15. NATIONAL SOCIETY FOR THE STUDY OF EDUCATION. *The Psychology of Learning.* Forty-First Yearbook, Part II. Chicago: Distributed by the University of Chicago Press, 1942.
16. PALMER, B. W. "School Situations Vitalize Mathematics." *The Mathematics Teacher* 37: 373–74; December 1944.

17. PLANK, EMMA N. "Observations on Attitudes of Young Children Toward Mathematics." *The Mathematics Teacher* 43: 252–63; October 1950.
18. "Professional Opportunities in Mathematics." *The American Mathematical Monthly* 58: 1–24; January 1951.
19. SAWIN, E. I. "Motivation in Mathematics: Its Theoretical Basis, Measurement, and Relationships with Other Factors." *The Mathematics Teacher* 44: 471–18; 471–78; November 1951.
20. SAWIN, E. I. *Motivation of Secondary School Mathematics.* Master's thesis. Department of Education, University of Chicago, May 1948.
21. STROUD, J. B. *Psychology in Education.* New York: Longmans, Green and Co., 1946.
22. TURNER, C. T. "Motivating Junior High School Mathematics." *Elementary School Journal* 36: 731–32; June 1936.

3. The Formation of Concepts

Henry Van Engen

INTRODUCTORY STATEMENT

EVEN a superficial study of the development of concepts uncovers the existence of a closely knit set of terms which are somehow or other almost welded together. It is not possible to dip into the field ever so lightly without encountering such terms as "meaning," "abstraction," "generalization," "learning," "understanding," and "perception," to mention only a few. There is of course a good reason for this because these terms refer to mental constructs which in themselves are concepts, and these concepts are essential for any intelligent discussion of the nature of concepts or the formation of concepts.

In addition to being impressed by this ever recurring set of words one is conscious of a fuzzy use of words, at least in some instances. If there is no fuzziness there is no common agreement as to the precise use of some of the key words. For example, Woodworth (53) speaks of "induction or concept formation." This use of the word "induction" may come as a surprise to mathematics teachers. Smoke (42) speaks of "concept formation, generalization, or concept learning." Other authors do not use the term "generalization" in this way but subordinate it to the term "concept" or "concept formation." Vinacke (48) in commenting on the fact that concept formation, as yet, is poorly understood says, "Thus, terms like 'abstraction' and 'generalization' are still utilized . . . without sufficient analysis of the behavioral and genetic processes involved" (p. 1).

The status of the term "meaning" is of particular significance at this time because of the emphasis it is receiving in discussions in the teaching of mathematics. Currently it is popular to write about meaning in arithmetic. Hence, many articles are appearing in current magazines about teaching arithmetic, and mathematics in general, meaningfully. Books on methods, and textbooks, must recognize this trend and discuss the topic at some length. The result is much confusion about "meaning" and some doubt as to whether everybody is talking about the same thing. This is especially true if one investigates the use of this term in those studies which deal predominately with the development of arithmetical concepts, although there are outstanding exceptions (See Werner, 52). Such terms as "meaning," "relationships," "understanding" are frequently used interchangeably as though they were synonymous. Hendrix (24) called attention to this problem in *The Mathematics Teacher* for November 1950. She pointed out that the terms "meaning" and "understanding" are not synonyms and, hence, to use them interchangeably in the discussion of concept development only serves to block the communication processes. So, in order to facilitate the communication of ideas it would seem best to clarify first the use of the terms "meaning" and "understanding"—at least, to designate how these terms will be used in this chapter of the yearbook.

THE MEANING OF "MEANING" IN MATHEMATICS

A word or a symbol is not always used in the same way. A symbol may have a meaning according to the way it is used in relation to other words, or according to how it is used in relation to objects, or according to the purpose for which it is used. Accordingly, it is customary to speak of the "dimensions of meaning." They are:

1. *The Syntactic Dimension.* Words and symbols have meaning because of the way in which they are used in relation to the other words in a sentence or formula. For example, "wind" means one thing in the sentence, "Wind the clock," and another thing in the sentence, "The wind blew hard." Similarly the symbol "2" means one thing in the equation "$x^2 + 5 = 6$" and yet another thing in the equation, "$x_2 + 5 = 6$."

2. *The Pragmatic Dimension.* Words and symbols will vary in meaning according to their purpose and consequences so far as a particular individual or organization is concerned. The words "economic royalists," "capitalists," and "socialists" as used in a political speech are likely to be so used in order to arouse the emotions of the listeners. Propaganda, both good and bad, is usually filled with words which function in their "pragmatic dimension." While this particular use of words is of little interest to the mathe-

matics teacher, as a mathematician, it should be of particular interest to him as an educator. The elementary teacher who punishes her pupils by making them work a certain number of arithmetic problems per offense is setting the stage for the future "pragmatic" use of the term "arithmetic." In such instances the term "arithmetic" is likely to become loaded with emotional overtones which will effectively block learning. Experiences in and out of the classroom take on meanings for the pupil, which meanings become associated with symbols. These meanings may be used later for a specific purpose by the child or adult. Witness the "big bad man," "the dark room" and the "ghost" as used at times by unwise parents. Mathematics as mathematics does not use words or symbols in the pragmatic sense but the "fringe" meanings of the symbol "mathematics" for the pupil is of supreme interest to the teacher as an educator.

3. *The Semantic Dimension.* This is the third dimension of meaning of primary importance to the teacher of mathematics, whether an elementary teacher of mathematics, or a secondary teacher of mathematics. Of course, when the child learns to combine symbols to express ideas he is employing words in the syntactical dimension. However, any sensible theory of instruction would insist that the child first learn that the individual symbols represent objects, other symbols, simple events, or mental constructs. These objects, or symbols, are called the referents of the given word or given symbol. A father points to a chair and makes the noise "chair." The child then soon learns that the object with four legs, a back and a flat surface, on which one sits, is the referent for the noise "chair." In this case the referent is a concrete object. It could be a group of objects such as "a herd of cows" or $x\ x\ x\ x$, the latter being a referent for the symbol "4 x's."

The referents may not be objects but may be "that which an object is doing." In this case the referent may be an action of a particular type. Thus, in the sentence, "The bird flies," the word "flies" refers to what the bird is doing—it refers to an action. For children a whole class of words and phrases, such as, "joining," "all together," "ran to meet," etc., refer to the action of combining two or more groups of objects which action is later symbolized by "+." Even symbols such as "+," "−," and "×," must have referents in the initial stages of instruction if arithmetic is to be taught meaningfully.

But referents may be abstractions, such as the words which represent colors, love, hate, or greed. Then again referents may be mental constructs, such as "mathematical proof," "variable," "function," "$\sqrt{-1}$," and many others readily recalled by mathematics teachers. The difference between these classes of referents is important to remember. In the one case the pupil derives the meaning of the word from sensory experiences. He actually hears, sees, smells, or feels the referent. In the other case the referent is a mental construct which may have its origins in sensory experience but these origins are not specifically identifiable. The origins of the sensory experiences which helped the pupil develop the mental construct for "$\sqrt{-1}$" are those experiences which developed the concepts for such numbers as 2, 5, −6, together with the concept of a mathematical system and a mathematical operation.

As has been stated previously, the teacher of mathematics is especially interested in the syntactic and the semantic dimensions of meaning. In the one case she is interested in teaching the child by means of symbols which derive their meaning from the way in which they are used in connection with other symbols. For example, the symbol "=" as used in the equation $3 + 4 = 7$. In the latter case, she is interested in making very clear to the child the referent for a symbol. For example, in the number 23 the 2 represents two groups of 10 objects while the 3 represents 3 of the same kind of objects. The referent then becomes the 2 bundles of 10 items and the 3 single items. Or to take a more abstract example, the algebra teacher interested in meanings is always very sure that when using the formula, $C = 3n$ (total cost of n three-cent postage stamps), the pupil knows that the n represents any positive integer and *only a positive integer* and that, consequently, the C represents *only* those positive integers divisible by three. That is, the teacher is making sure that the pupil knows the referent for the symbols used in the formula. Such a teacher will never let the child draw, in this instance, the usual straight-line graph for the formula $C = 3n$.

THE FORMATION OF CONCEPTS

There can be little doubt that in general the teacher of arithmetic and the teacher of mathematics spend a disproportionate amount of time on the syntactic dimension of meaning to the neglect of the semantic dimension. Mathematics is too often taught by drill on the mechanical features of mathematics. This places the emphasis on the syntactic dimension of meaning. Before the pupil is ready to use a symbol in connection with other symbols the teacher must have established a referent for the symbol in isolation. For example, in arithmetic, why drill on the combinations (syntactic dimension) before making sure that the child knows the meaning of such symbols as 4, 8, and 9 (semantic dimension). What are the effects on the child when he is drilled on combinations (syntactic dimension) while the teacher totally neglects to establish the referents for such symbols as $+$ and $-$ (semantic dimension)?

Generally speaking, the teacher is confronted with the problem of establishing the semantic dimension of meaning before establishing the syntactic dimension. In many cases however it is not possible to cut up a teaching situation into three parts using the three aspects of meaning as a "knife." Syntactic and semantic dimensions can, and do, coalesce. However, the teacher of mathematics must keep in mind constantly that words when used in context do not have the same meaning as words when used in isolation. Having established the meaning of a word when used in isolation the teacher is still confronted with the problem of establishing the various shades of meaning of that word when used in context.

As an example of the shades of meaning which a symbol takes on when used in connection with other symbols, consider the use of the symbol x in the following situations: (a) $2x + 3$, (b) $2x + 3 = 5$, (c) $1/x$, (d) $(x + 2)^2 = x^2 + 4x + 4$, and (e) $(x + 2)^2 = x^2 + 2x + 4$. When considered from the point of view of the first semester course in algebra in the high school the x in (a) designates any number selected from the class of real numbers. In (b) the x represents any number selected from the class of numbers which has only member, namely, 1. Other restrictions can be made to fit examples (c), (d), and (e). In fact, in all cases except (a) and (d) the class of numbers from which numbers may be selected to be substituted for the variable x is a different class. Thus, the meaning of the symbol x in these examples varies from example to example because of the way it is used in connection with other symbols.

While it is true that the mathematics teacher, as a mathematics teacher, is not concerned with the pragmatic dimension of meaning, as an educator she must be greatly concerned with this dimension. Too many children, and adults, "freeze" when they think, or hear, the word "arithmetic." On the other hand, for some pupils the word "arithmetic" or "algebra" releases very desirable emotional responses. These responses are the result of experiences in the classroom, and the results of these experiences must be kept in mind by all teachers of mathematics.

A CLARIFICATION OF THE "MEANING" OF UNDERSTANDING

To discuss the "meaning" of understanding by reviewing all the various ways that the word "understanding" can be used would be of little value for the teacher of mathematics. The main problem centers around how the word may be used by children in the elementary- and secondary-schools. What are the methodological implications of remarks made by children and adolescents, such as "I do not understand it," "What do you mean?" and "I know what you mean but I do not understand it." These questions, when used in a mathematics classroom, have special significance. The pupil who asks, "What does this mean?" is searching for something different from that which is wanted when he says, "I do not understand it." Of course, the pupil's difficulty may be a combination of both "meaning" and "understanding" but for the sake of clarity it will be best to consider these problems separately.

From the way these two words are used it is evident that they can have different meanings. One speaks of "the meaning of a symbol" and not "the understanding of a symbol." Similarly, one says "I understand your proposition" and not, "I mean your proposition." ("I mean your proposition" is a statement designating a proposition which has been introduced into the discussion. It merely points to a given proposition.)

Understanding refers to something that is in the possession of

THE FORMATION OF CONCEPTS

an individual. The individual who understands is aware of a satisfying feeling, a psychological closure, which results from having fitted everything in its proper place. Of course, this psychological closure must be tested because a child may think he understands when he does not understand. This problem may be considered elsewhere, however.

The pupil who understands is in possession of the cause and effect relationships—the logical implications and the sequences of thought that unite two or more statements by means of the bonds of logic. The statement which is understood is seen to follow from statements accepted previously by the pupil. From this point of view it is seen that understanding implies an "if-then" relationship. The pupil who does not understand cannot explain how a given "if" implies a "then." The pupil who understands can explain the "then" by means of an "if." There are however, kinds and degrees of understanding. In this connection, see Hadamard (11).

Consider the following example:

$$\begin{array}{r} 56 \\ -18 \\ \hline \end{array}$$

The pupil, or adult, is to be taught how to work this example by the equal additions method. Whatever the words used to show how to do this example the import of the words will be that "If you add 10 to the minuend and subtrahend then you can more readily do the subtraction." Now after the pupil has been shown that the 10 is added to the 6 in the minuend and the 1 (ten) in the subtrahend he will know what the above statement (placed in quotes) means but unless he knows the additional mathematical principle that adding 10 to the minuend and subtrahend does not change the answer (difference) then he cannot understand why this procedure obtains the right answer. In other words he knows the meaning of the statement, "You can subtract if you add 10 to the minuend and subtrahend." but he does not understand it. Obviously, a pupil may also be able to perform this subtraction with great skill but he may **not** have reached that closure which is so essential for understanding.

Meaning is that which is "read into" a symbol by the pupil.

The pupil realizes that the symbol is a substitute for an object. It is a triadic relationship between a pupil, a symbol, and the referent. Understanding is more nearly a process of integrating concepts—placing them in a certain sequence according to a set of criteria. Meaning, in its semantic sense, is a substitution process. It is a substitution of symbol for object, or symbol for symbol or symbol for concept. Understanding is an organizational process.

From these considerations it would seem that the phrase, "I know what you mean but I do not understand it" is not a mere play of words. In a particular instance the pupil may know the referent: he may know what to do but he may not know why it should be done: he cannot make the logical connection between the "situational need" and the response.

The teacher of mathematics will teach for both understanding and meaning. This is a trite statement, of course. But what does a teacher do when she teaches for "understanding"? What does she do when she teaches for "meaning"? An example may help establish the essential difference in objectives in these two instances.

Assume the teacher is confronted with the task of teaching the theorem in geometry relating to the measurement of an inscribed angle. Now if she wishes to establish the meaning of the statement, "An inscribed angle is measured by half of its intercepted arc," she would make sure that the pupil knows that if the arc is 80° then the angle is 40°. For this purpose, visual aids of various kinds are very important. The pupil quickly forms various inscribed angles on the visual aid and in each case finds that the measure of the arc is twice the measure of the angle.

But at this point the pupil can still say, "I know what you mean but I don't understand it. Why is it so?" Now the teacher is confronted with the task of showing the pupil how this statement follows as a consequence of having accepted (and understood, let's assume) previous statements about central angles and circles. In other words, to help the pupil understand the statement, "An inscribed angle is measured by one-half of its intercepted arc," the teacher helps the pupil to fit it into a conceptual structure already in the pupil's possession. If the pupil does not have this conceptual structure it is obviously impossible for him to know why it is possible to make this statement about inscribed angles.

The teacher of mathematics should fully realize the difference between these two approaches. She should realize that certain methods are appropriate for the development of understanding and still other methods may be appropriate for the development of meaning. She should realize that the same methods are not necessarily appropriate for the development of both understanding and meaning. The teacher who makes these distinctions and adjusts her methods accordingly must of necessity be a better teacher than the one who blindly strives to "teach meaningfully."

THE NATURE OF A CONCEPT

In spite of the fact that the educational world has long been confronted with the problem of developing concepts, mathematical and otherwise, there is much that is not fully understood about the nature of a concept. This state of affairs is not due to a lack of interest in the problem or to a failure of intensive investigation. The problem of how children and adults develop the ability to form a concept is very complex. The problem is made more difficult by the fact that the processes whereby concepts are formed may not be the same for the child as it is for the adult. There is reason to believe that the perception-abstraction-generalization-response sequence, which seems to play such an active part in the development of the child's concepts, is not so predominant in the development of the concepts of the adult. On the other hand, one must not assume that because the problem is not completely understood conceptual development has no general features which are usually accepted by those who have given the problem considerable thought.

In view of the present knowledge of the nature of a concept it would probably not be wise to consider the pros and cons of the various definitions that can be found in the literature. Compare, as examples, Harriman (12), Smoke (41), Warren (49). Instead it would seem wise to discuss the general features which are covered by the various definitions following which a summarizing statement about the nature of concepts can be given. Such a procedure should prove to be more useful to the teacher of mathematics in both the elementary and secondary schools than a concise, formal definition of a particular point of view regarding the nature of a concept.

1. While it is true that concepts are not sensory data, yet they are "that something" which results from numerous sensory experiences which are combined, generalized and carefully developed. The argument that this situation obtains can be supported both by bringing logic to bear upon the situation and by citing common classroom experiences.

The child, seemingly, is born with the innate ability to develop concepts. Sensory experiences are essential to awaken this innate ability so that conceptual development can begin. The child entirely shut off from the world would not develop the concept of "dog" because the sensory experiences are lacking out of which this concept must arise. If he cannot feel a dog, see a dog, and hear a dog he does not have the fundamental units from which to construct a concept of "dog."

Experimentally, it is easy to demonstrate, as Vinacke (48: 3) shows, that concepts are an elaboration of sensory data; however citing classroom experiences rather than experimental results may serve a better purpose. Every thinking teacher has experienced situations in which a child is blocked from responding because of a lack of sensory experience—usually referred to under the broader heading of experience. Of course, from an experimental standpoint such situations are crude examples but they serve to make the point. Children fail to distinguish between colors, in part at least, because they have not had sufficient experience, or an experience, with a particular color, or particular colors. In fact (except for color blindness), for many children it would be impossible to pick out the color magenta for one of the following reasons: (a) they have not experienced—have not seen—the color magenta, (b) they may have experienced the color but have not associated the word "magenta" with the sight experience, or (c) they have not made or consciously compared the color sensation produced by a magenta color with other colors. In any case the child is blocked from making correct responses (ignoring the chance factor in this instance).

Teachers of reading have long recognized the importance of a wide background of experience because children cannot talk about or read about concepts which they do not possess. This wide background of experience is nothing more than a large collection of sensory experiences which are in the possession of the child and available for recall upon the receipt of the proper stimulus. From this it follows immediately that the concepts possessed by the child will depend upon the previous experiences of the child. Furthermore, in view of the fact that no two children have the same experiences, in toto, it follows that the concept "dog" possessed by individual A is not the same as the concept "dog" possessed by individual B. As an example, consider two geometry students, P and Q. Student P has drawn similar triangles and noticed that any two pairs of lines in one triangle have the same ratio as the corresponding pairs of lines in the other triangle. Student Q has only noticed that the corresponding sides of similar triangles have a constant ratio. Student P will have a different and more complete concept of similar triangles than student Q. Why? Because student P has had a different experience with similar triangles than has student Q. Under formal teaching this difference in experience may have been a difference in a "definitional experience." That is, similar triangles may have been defined as being similar if, and only if, any two pairs of lines in one triangle are proportional to the corresponding pairs of lines in the other triangle rather than the definitions found in the usual geometry classroom. But regardless of the approach used the stimulus of the spoken word "triangle" will be associated with a richer background of experience in one case than it will in the more limited experiential situation; and hence, the concept of "triangle" will differ in each case.

2. Children and adults tend to integrate sense impressions and respond to various stimuli in the same way. This feature of concepts is readily illustrated by those individuals who will react to the shouted word "mouse" in much the same way that they react to the sight of a mouse. The geometry student reacts to the s.a.s. condition in the same way that he reacts to the a.s.a. condition in spite of the fact that the stimulus is different. In this case the two stimuli have been thoroughly integrated so as to

elicit the same response—"congruent." In much the same way the pupil integrates the sense impressions received by observing variously shaped triangles with all the known (by him) properties of a triangle and responds to them with the thought, "that's true about all triangles."

3. The integration of sensory experiences, discussed above, is accomplished through a symbolic process; at least it is symbolic in nature. In the case of the human organism this integration is usually accomplished by means of words. Different words (symbols) in many instances elicit the same response from the organism. Many simple instances of the integration of sensory experiences to form a concept can readily be given. The word "car" ties together the sensory experiences of the adults of today in many ways. The pleasurable experience of the Sunday afternoon ride, the experience of the near accident, or the injuries received in an accident, are combined to form a total concept of "car." The type of overt or emotional response released by the word "car" will depend on the particular integrated set of experiences which are symbolized by the word "car"—in other words, the concept of "car."

Mathematical instances of the symbolic integration of various sensory experiences which are essential to the formation of a concept can be readily given. The child in his early work in the elementary school has many experiences with the actions associated with taking $\frac{2}{3}$ of some physical object such as a sheet of paper. Later this same symbol may bring forth a different response—that of taking $\frac{2}{3}$ of a number of objects. In this case the symbol ties together different, but related, sensory experiences.

4. There can be no doubt that sensory experiences play an important part in the development of the primitive concepts of an organism. On the other hand, for certain very abstract concepts the sensory experiences are so far removed, or so intangible, that the sensory origins of some concepts are hard to locate. In fact, from an introspective point of view there is reason to believe that the more advanced concepts of the human organism do not depend on sensory experiences. Such concepts as "proof" in mathematics, or the concept of a mathematical system are at least far removed from sensory experiences. The concept of "im-

plication" as used in mathematics would seem to result from an insight into how words are used rather than from sensory experiences. This point of view is supported according to Woodworth (53) and Munn (32) by what is known about the debatable subject of imageless thought.

5. When one considers what happens inside the organism which might help characterize a concept there are two considerations which seem important.

a. Concepts represent selective mechanisms—a "sieve" through which external stimuli must pass in order to arouse symbolic responses in the pupil. Also the reverse situation may obtain, i.e. the symbolic response arouses a perceptual response. Thus the child may see the three-sided figure on the blackboard, thereby arousing the response "triangle," a response which was selected from the myriads of other responses that could have been given, or he may hear the word triangle and look up to the board to see a triangle.

b. While experiments indicate this selective mechanism as a characteristic of concepts it should not be inferred that a verbal, or symbolic response is a necessary condition. Hendrix (25) very ably makes this point and Smoke (43) obtained experimental evidence which suggests that individuals may possess a concept and yet not be able to give evidence of the possession of this concept by means of the usual lines of communication. Smoke's (42) definition of concept learning recognizes the existence of non-symbolic responses. Hebb (14) says, "The implication of the preceding paragraph is that a concept is not unitary. Its content may vary from one time to another, except for a central core whose activity may dominate in arousing the system as a whole. To this concomitant core, in man, a verbal tag can be attached; but the tag is not essential" (p. 133).

The existence of concepts on the subverbal level is a condition that should be of utmost importance to the teacher of mathematics. Children often give evidence of being in possession of well-defined number concepts which enable them to solve numerical situations, but being unable to tell how they solved the problem. Upon being pressed for an explanation the child, at times, reacts by saying, "It's just so," or responds with the not too complimentary statement, if pressed too hard, "Ain't he dumb?" In teaching, Hendrix (25) points out, it is important that this awareness of a generalization be developed before it is given a name. Many times teaching proceeds by naming a generalization and then giving instances for the purpose of making the pupils aware of the generalization.

A summary statement about the nature of concepts is now in order. For this purpose one can do no better than let Vinacke (48) do the summarizing. However, it should be noticed that Vinacke does not, seemingly, recognize the possession of a subverbal concept by an organism in his summary.

> They (concepts) must be regarded as selective mechanisms in the mental organization of the individual, tying together sensory impressions, thus aiding in the identification and classification of objects. But concepts involve more than the integration of sense impressions, against the background of which recognition occurs, for they are linked with symbolic responses which may be activated without the physical presence of external objects. That is, concepts can be given names—can be detached from specific instances, by means of a word—and used to manipulate experience over and beyond the more simple recognition function. The symbolic response, however, stands for whatever it has been linked within the previous experience of the organism and depends upon how that past experience is organized (p. 5).

THE ATTAINMENT OF CONCEPTS

The development of concepts is basic to growth in learning capacity. In general this growth in learning capacity is a growth in conceptual development. For this reason it is important to make a study of how concepts are formed and to make an application of this knowledge to the methods employed in the classroom. Any such study must of necessity consider the activities of abstracting and generalizing, since these activities are inevitably a part of the total process of concept formation.

When a pupil observes common sensory or perceptual qualities in a number of different situations, or objects, he is abstracting that quality from the total situation. The concept of "green" is acquired by seeing green in connection with many different objects and colors and then focusing the attention on the one

THE FORMATION OF CONCEPTS 83

element in common—namely, green. Green then eventually is thought of as a "thing" in itself, something separate and apart from everything else. The concept of a number may be acquired in much the same way. The child may attach the number five, for example, to a particular group such as the five fingers of the hand. Later he observes that "five" applies to other groups of objects as well. Eventually the child abstracts a "fiveness" which is common to those groups which can be put into one-to-one correspondence with the fingers of his hand. It is well known that there are primitive tribes which have not made this abstraction. They apply different number names to different groups equal in number but differing in the kind of objects which make up the group.

Abstraction plays an important role in the classification of objects. A common property is fixed upon as the criterion for including an object in a given group. Each item is then examined to see if it should be included in the group. For example: If numbers are to be classified as prime or not prime each number is examined to see if it is divisible by a number other than itself and one. If it fails to pass this test it is classified as a prime number, otherwise it is not a prime. Here a certain type of divisibility was the abstraction which determined how each number was to be classified.

Generalization is another process used in conceptual learning. Generalization signifies that the detail which has been abstracted from a group of objects, or situations, is used to respond similarly to a whole class of related objects or situations. Thus, a student who understands the Pythagorean Theorem has abstracted a property common to all squares constructed on the hypotenuse and the legs of a right triangle. This property can be used, however, to respond similarly to a whole class of other situations. If similar polygons are constructed on the hypotenuse and the legs of a right triangle the same abstraction can be applied to the polygons as was applied to the squares in the Pythagorean Theorem. On the other hand, this "Pythagorean property" can be generalized to a much larger class of objects. Any figure, curvilinear or otherwise, constructed on the hypotenuse of a right triangle will have its area equal to the sum of the areas of the similar figures constructed on the two legs of the triangle. Thus a pupil discovers that an abstraction which he has learned by considering a rather limited case (the squares) also covers innumerable other cases.

How are these two processes of generalization and abstraction used in explaining the formations of concepts? Two theories are commonly recognized. The one emphasizes the passive role of the individual and abstraction, while the other emphasizes the active role of the individual and generalization. The theory emphasizing the passive role of the individual and abstraction can best be stated by quoting directly from Hull's (27) classical experiment on concept formation.

A young child finds himself in a certain situation, reacts to it by approach say, and hears it called "dog." After an indeterminate intervening period he finds himself in a somewhat different situation and hears that called "dog." Later he finds himself in a somewhat different situation still, and hears that called "dog" also. Thus the process continues. The "dog" experiences appear at irregular intervals. The appearances are thus unanticipated. They appear with no obvious label as to their essential nature. This precipitates at each new appearance a more or less acute *problem* as to the proper reaction . . . ; the intervals between the "dog" experiences are filled with all sorts of other absorbing experiences which are contributing to the formation of other concepts. At length the time arrives when the child has a "meaning" for the word dog. Upon examination this "meaning" is found to be actually a characteristic more or less common to all dogs and not common to cats, dolls, and teddy-bears. But to the child the process of arriving at this meaning or concept has been largely unconscious. . . . Such in brief is our standard or normal-type of concept evolution (p. 5).

The active role of the individual and generalization is emphasized in the theory of concept formation which stresses that the concept originates as a hypothesis which is tested by applying it to members, or supposed members, of a class of objects or situations.

While these two theories stress different features in the concept formation process, actually it seems better to unite the two to form an eclectic theory of concept formation. This course is indicated because it is difficult to distinguish between generalization and abstraction in the actual behavior of an individual. Heibreder's (23) experiments have shown that in conceptual learning both processes operate at the adult level. However, there is reason to believe that for the child the sequence of perception-abstraction-generalization is more nearly a true statement of affairs than it is for the adult.

Heidbreder (15) in studying concept formation has achieved a number of very significant results showing the order in which adults attain certain types of concepts. According to Heidbreder those concepts were attained first in which the abstractions could be made by reacting to drawings of pictured objects of things such as trees, faces, and buildings. Next in difficulty she found that abstractions could be made by reacting to drawings of forms— "something less than a thing but not altogether un-thing-like." Such forms as circles, squares and triangles were used in her experiments in this instance. Most difficult of all the concepts studied were those in which the abstractions had to be attained by reacting to facts about collections of objects (numerical quantities of). This latter type of response seems to be more remote from the perception of concrete objects than is the response to such things as visual forms and spatial forms.

What explanation does Heidbreder (23) give for this order of things-form-number concept attainment? Using her own words, "One answer immediately suggests itself: manipulability, relevance to direct motor reaction" (p. 182). The concept "circle" is attained significantly later than the concept "plate" because a circle is beyond the manipulability stage even though it is as perceptible as the plate. The plate can be manipulated with the hands, it can be felt, seen and weighed. The circle drawn on the board cannot be manipulated. Its form can be traced in the air but even this cannot offset the advantage of manipulability from the standpoint of the attainment of the concept. The ease of attaining a concept seems to be more highly correlated with manipulability than with perceptibility. In Heidbreder's own words:

> Dominance in cognitive reactions seems to be correlated, not with maximal openness to inspection, nor with maximal "givenness" in perceptual experience, but with *maximal relevance to action, specifically to manipulation*, that kind of motor reaction which human beings characteristically employ (p. 182) [italics added].

Reactions to the world of concrete objects are the foundation stones from which the structure of abstract ideas arises. These reactions are refined, reorganized and integrated so that they become even more useful and even more powerful than the original response. The part of the reorganization and the refinement of responses in conceptual learning is nicely described by McConnell (31). Further consideration cannot be given to this problem at this time.

This concept of the part that actions, or manipulations, play in the development of concepts of the first order is of utmost importance to the teacher. Children and adolescents use manipulatory experiences to develop primtive concepts, and those concepts which are more nearly related to the action world of the child are the ones that are more easily developed. From this it would seem that any effort to improve the instruction in mathematics must take into consideration this rather commonly accepted point of view regarding the attainment of concepts. The weakness of mathematical instruction as commonly practiced in our schools is more readily observable in respect to the lack of adequate activities for conceptual development than in almost any other respect. Books, paper, pencils, blackboards, and the drill exercises which usually accompany these instructional tools, are not sufficient except, possibly, for that relatively small percentage of pupils who are symbolically minded. Today's schoolrooms are barren of those small inexpensive objects which provide those opportunities for perceptual and manipulatory experiences from which the average child can abstract and generalize in order to take the first steps in formulating a concept. This barrenness is further accentuated by the lack of pictures, movies and filmstrips which can be used to picture the concept-forming actions as a second stage in the learning of abstract symbolism. This action-picture-symbol sequence in concept formation is frequently ignored in its entirety and the symbol introduced immediately. No other method can so thoroughly block conceptual learning, especially for the average and the slow-learning pupil.

The present day emphasis on multi-sensory aids has a secure foundation in the point of view expressed by Heidbreder, Smoke and other psychologists of the operationalists' school of thought.

THE FORMATION OF CONCEPTS

The manipulatory activities that result when blocks, buttons, and models are widely used in instructional procedures provide the essential elements from which concepts are more readily developed. However, this emphasis on multi-sensory aids does place a responsibility on the teacher of mathematics which must be given careful thought. Many of the manipulatory activities now "going the rounds" in the world of mathematics instruction do not include those manipulatory activities which develop the concept, or concepts, for which they were developed. As textbooks, workbooks, and tests need careful evaluation, so also do visual aids. As one example of a visual aid which does not aid the child in developing the desired concept the following may be cited as a horrible example.

Sometimes one finds that the first-graders are being taught "what subtraction means" by the following picture device.

$$OOOOO - OO = OOO$$

Now any teacher knows (and the child knows this even better than the teacher, seemingly) that one cannot actually perform the operation pictured above. It is impossible to take two marbles from five marbles as illustrated. Laying five marbles on the table and giving two of them away will easily establish this fact. Now the child also knows this because he has performed the feat of giving two of his five marbles to Johnny many times. As a result the "learning aid" illustrated above can only block the real meaning of subtraction for the child in the first grade. It blocks learning because it does not picture the real life actions that indicate to the child what the word "subtraction" means.

NUMBER-CONCEPT FORMATION IN CHILDREN

Previous sections of this chapter have already commented on the fact that the difference between the processes whereby adults form concepts and the processes whereby children form concepts are likely to be one of degree rather than of kind. Therefore, the action-manipulatory point of view set forth by Heidbreder as quoted in the last section of this chapter are fully as appropriate for the processes of concept formation in the child as they are in the adult. In fact, there is reason to believe that they are even more characteristic of the child's conceptual processes than they are of the learning processes of the adult. It will be the purpose of this section to support the point of view that actions and manipulations are dominant in the formation of the child's concepts.

There can be no doubt that too little thought has been given to the part that actions play in the intellectual development of the child. Attention has already been called by Van Engen (47) to the role of actions in the intellectual development of the child when this development is considered from a philosophic and semantic point of view. However, the psychological foundations of the "action basis" for learning must also be given serious consideration. This is forcefully brought to mind by the phrase "human action system" which Gesell (10) used in his study of the growth aspects of the mind. For Gesell, "action system" denotes "the total organism as a going concern, particularly its behavior capacities, propensities, and patterns."

To those teachers who have thought of learning in terms of "specific habits" "drill" and "teaching by telling" it comes as somewhat of a shock to learn that probably all mental life has at its roots the actions or manipulations performed in a learning situation. Gesell (10) makes this point as follows:

> It is probable that all mental life has a motor basis and a motor origin. The non-mystical mind must always *take hold*. Even in the rarefied realms of conceptual reasoning we speak of intellectual grasp and of symbolic apprehension. Thinking might be defined as a *comprehension* and *manipulation* of meanings. Accordingly, thought has its beginnings in infancy. We have already noted the germ of mathematics which lies in the one-by-one behavior pattern of the year-old infant. Counting is based on serial motor manipulations (p. 58).

The principle of action is tightly interwoven into Gesell's description of the growth processes of the child (mental, physical and emotional). For example, he says:

> This principle [of motor priority] is so fundamental that virtually all behavior ontogenetically has a motor origin and aspect. Vision, for ex-

ample, has a motor as well as a sensory basis; likewise speech, mental imagery, and conceptual thought. Even emotions trace to motor attitudes and tensions (p. 65).

These quotations taken from the latest of Gesell's publications furnish much food for thought for those interested in the contributions of classroom experiences to the growth of mathematical concepts. But before considering these implications in some detail it will be profitable to consider the point of view of others who have given this problem considerable thought.

The next source from which quotations will be taken is not strictly a study on concept formation. It is rather a study on the development of reasoning in the child; yet thinking is merely the mental manipulation of symbols which represent concepts. The close relationship existing between the mental manipulation of symbols and the overt manipulation of objects is apparent upon reading Piaget's (35) classical studies on reasoning and judgment in the child in the light of the discussion found in the previous sections of this chapter. Although later studies have failed to corroborate Piaget's studies in all details, it is nevertheless a pioneering effort which set the stage for many investigations and can be relied on for its basic point of view.

Piaget emphasizes the close connection between manual operations, or actions, and the thought processes of the child. In fact he holds that the child thinks by picturing, mentally, the manual operations that took place in a given situation. Thus, Piaget (35) says,

So that everything we have said in this work is to show that the thought of the child is less conscious than ours has *ipso facto* led us to the conclusion that childish thought is devoid of logical necessity and genuine implication; *it is nearer to action than ours, and consists simply of mentally pictured manual operations*, which, like the vagaries of movement, follow each other without any necessary succession. This will explain later on why childish reasoning is neither deductive nor inductive; it consists in mental experiments which are non-reversible, i.e., which are not entirely logical. . . . (p. 145–146) [Italics added].

These movements and operations are a preparation for conscious reasoning in so far as they reproduce and prepare anew the *manual operations of which thought is a continuation* (p. 145) [Italics added].

This is strong language; it has many important teaching implications. Can it be supported? Werner (52) says,

The child's concepts always have a concrete content. Image and concept are an indivisible unity. The conceiving and the describing of a thing are not distinctly separated activities. As is true of primitive man, the child's need of adjustment to adult language creates conceptual forms which arise out of concrete perception, which are indeed both perception and conception, which appear to be metaphors and yet really are not.... To conceive and define things in terms of concrete activity is in complete accordance with the world-of-action characteristic of the child (p. 271–72).

Here again one finds the action and the place of action in the development of the concepts of the child. Are number concepts developed this way? Consider the following quotations also taken from Werner (52).

Frequently we find that abstract counting is supplanted by an optical, or even motor configuration and ordering of groups among primitive peoples and, indeed, among the *naïve* of our own culture.

The formation of a 'number system' in its proper sense is bound up with two developmental facts: First, with the increasing abstraction; the number concept becomes more and more released from the concrete configuration and the qualities of the objects. Second, with the development of a scheme for the number order in particular (p. 294).

In view of the importance of the topic and in view of the importance of the contribution to the field it may be well to quote Heidbreder's (17) findings and her interpretation of these findings from a source not previously quoted.

Definitions referring to concepts of number were especially instructive. ...

There are thus indications that in attaining concepts of numbers, some subjects reacted first chiefly to pictured objects, next chiefly to spatial arrangements, and eventually chiefly to numerical quantities, thus traversing in arriving at these, the last concepts attained, the entire course of events indicated by the experimental data considered as a whole.

Taken together, the quantitative data and the definitions are interpreted as indicating that in successive stages of an experiment, the subject's reactions were critically determined by successively less thing-like aspects of the drawings as reaction determined by more thing-like aspects proved inadequate (p. 137).

While Heidbreder's experiments dealt with the attainment of number concepts in adults its significance for the development of the child's concept of numbers cannot be overlooked. The response to a spatial arrangement prior to the response to the numericalness of a situation is particularly instructive. It would seem to indicate that configurations play a fundamental role in the development of number concepts.

Judd (28) points out that

Number ideas are, in fact, more than images; they depend on the presence of reactions. A child does not learn numbers by having them impressed on his organs of sense. There is no such thing as a number sense. Number is acquired only when there is a positive reaction. One must respond in a definite way to each item of experience which is to be counted. The definite positive response which one makes to each object counted is reduced to an inner reaction in the course of educational development, but it continues to be a reaction (p. 49).

From these quotations one can conclude that the perceptual, manipulatory activities are of utmost importance in the development of number concepts as well as concepts in general. On this basis one can again conclude that in this respect instructional practices in the elementary school are in general very weak. The usual blackboard-chalk-paper-pencil methods for instructing the child in arithmetic are entirely inadequate. Making marks in a workbook is not a functional activity in the *first* stages of concept development. Neither is continual drill on abstract combinations of symbols functional. These quotations show, clearly, that the manipulation of objects is essential in the *first* stages of number concept development—especially in children. Abstract definitional approaches should be abandoned by the elementary teacher and secondary teacher for an approach which emphasizes the organic awareness of a concept before it is characterized by a definition or designated by a symbol.

The effect of this hypothesis pertaining to conceptual learning is nicely brought out by present-day practices in teaching children to count. Too prevalent is the idea that the first stage in counting is the memorization of a sequence of number names. Nothing could be more erroneous. The part that "number configurations" play in the development of the ability to count has not been thoroughly investigated but there can be no doubt that a configurational awareness of number should precede the sheer memorization of number names.

IMPLICATIONS FOR THE TEACHING OF MATHEMATICS

What is known about concept formation and the implications for mathematics teachers can probably be best, and most economically, set forth in a few statements which, in part, summarize what has been discussed in earlier sections and, in part, a brief statement of the results of experimental evidence not previously discussed. When new results are included in the following statements a reference to the bibliography will be given.

1. Mathematics teachers have not made enough use of what Heidbreder calls the "thing-like" aspects of conceptual learning. The initial experiences with a new concept should conform to the "world-of-action" characteristic of the pupils conceptual learning processes. The ease with which concepts are acquired depends to a great extent on the "relevance to direct motor reaction."

2. There seems to be evidence that the more intelligent the pupils, the more they are able to deal with language symbols, and that they rely more and more on such symbols as the problems become complex (8). The fact is not surprising but it has particular significance for the teacher of the slow learner. These pupils are weak in the use of symbols and yet the instructional tools placed in the hands of the pupil deal almost exclusively with language symbols. Concrete-action learning equipment is needed in any attempt to solve the problem of the slow learner. The slow learner needs the manipulating experiences which develop concepts. He also needs picture sequences to encourage him to become independent of the concrete learning aids.

3. Verbal instructions increase the variability of response (8). Teachers should be conscious of the fact that pupils interpret the meaning of words in terms of their own experiences and that these experiences are not the same as those of the teacher. Hence, visual aids will help to "unclog" communication lines, avoid misunderstanding and decrease the variability of response.

4. "A combination of abstract presentation and concrete examples yields a distinctly greater functional efficiency than either method alone" (27).

5. "During the evolution of concepts, mildly attracting attention to the common element 'in situ' considerably increases the efficiency of the (learning) process" (27).

6. "A set to learn meanings as well as names yields a much higher rate of learning and degree of retention than a set to learn names only" (36).

7. Concepts logically learned are learned more quickly and are remembered longer than are concepts illogically learned (36). Commenting on this fact Stroud (45) says,

Material high in associative value is for that reason comparatively easy to learn and for the same reason easily recalled, relearned or recognized afterward. Logical material, material capable of meaningful organization or reduction to some kind of system, comes within the operations of transfer of training, operations that facilitate recall as well as learning (p. 538).

8. A given situation will favor one concept over another and the ease of attainment of the concept will depend on how readily discernible the essential features of the concept are to perception (48).

Teachers might well keep this generalization in mind in evaluating the visual aids used in their classes. Too many visual aids in use today do not highlight the essential features of the concept they are supposed to teach. In many cases the essential features are too imbedded in the total situation. In still others it is merely a visual aid, there is no relevance to the development of the concept.

A simple example may make this clear. Teachers frequently provide counting experiences in which the actions essential for establishing the cardinal concept of number are not readily accessible to perception. This inaccessibility to perception frequently causes the pupil to confuse the ordinal and cardinal concept. If teachers ask the pupils to count six children in a row the predominant features in this situation are essentially ordinal not cardinal. However, in counting six blocks which can be put into familiar number configurations successively, each group is grasped as the child says, "One, two, three, four, five, six." All the actions here present facilitate the development of the cardinal concept, and the eye is aided in seeing the total group and not each successive member of the group as an individual.

9. Negative instances are not necessary for the development of adequate concepts but may be included as checks (42).

In teaching the concept of adjacent angles in geometry the teacher may include drawings of angles which have a common side but not a common vertex. This is a negative instance inasmuch as it does not fit the definition of adjacent angles. However, research has shown that the inclusion of negative instances does not materially affect the development of the concept.

10. Conceptual development is a growth process. It takes time to develop concepts. Hence the teacher should not expect the pupil to develop a mature concept in a few days. Concepts are developed by reviewing various instances in which they may occur under varying conditions and with varied meanings. Furthermore, concepts are not established readily by definitions unless the pupil is mathematically mature.

11. There are such things as nonverbalizable generalizations. Hendrix (25) and Smoke (43) have discussed the existence of nonverbal generalizations. Hence the pupil who says, "I know what it is but I cannot say it," may be telling the truth. Furthermore developing the awareness of a generalization prior to verbalization facilitates learning.

12. The background of experience of the pupil is an important factor in the development of concepts. This is amply illustrated by the pupil who has done some art work and knows the term "perspective" as it applies to representing three-dimensional objects on two-dimensional paper. This student when confronted

with the term "perspective" as used in college geometry is often confused. He is looking for the third dimension and there is no third dimension.

The child who has had many and varied experience with money encounters much less difficulty with the "arithmetic of money" in the primary grades than the child whose experiences have been limited. Similarly the child who has been given many varied experiences with number ideas will have less difficulty with the abstractions presented at a later date.

From this point of view it would seem that it is the teacher's duty to give the children many varied experiences with the concrete objects and the manipulations and actions which are essential to the development of concepts. Such activities are essential to good instruction in mathematics.

13. Conceptual thinking is not necessarily harder than concrete thinking but it is easier to manipulate the concept of Sam Jones than it is of Mr. A or of a. Murphy (33) makes this point very nicely as follows:

> The abstraction "man" or "Mr. A" is actually handled less efficiently in logical relations than is Mr. Edward Jones or Mr. Harold Smith. The results of experiments in reasoning in which the same rational processes must be carried through first with concrete and then with abstract materials show gross differences. Reasoning depends not on the formal ability to take the necessary logical steps, since the connections required to solve the task are the same in the two cases. But sometimes, in thinking in terms of abstract things like x's and y's, or Mr. A and Mr. B, one is unable to control the concepts and handle them in pure form. One is trying to do two things at once—concrete and abstract. In the concrete tasks one manipulates spatial relations in pictorial or other form which keeps them in the realm of immediate experience rather than abstraction (p. 391).

Algebra teachers should keep this in mind. The beginning algebra student has learned to think in terms of individual numbers such as 3.14, but he may have difficulty in thinking about an abstract symbol which represents that number, such as π. In particular, symbol x represents a class of numbers which may cause considerable difficulty.

Bibliography

1. BROWNELL, W. A., and HENDRICKSEN, G. "How Children Learn Information, Concepts, and Generalization." *Learning and Instruction.* Forty-Ninth Yearbook, Part I. National Society for the Study of Education. Chicago: University of Chicago, 1950. p. 92–128.
2. BROWNELL, W. A., and SIMS, V. M. "The Nature of Understanding." *The Measurement of Understanding.* Forty-Fifth Yearbook, Part I. National Society for the Study of Education. Chicago: University of Chicago, 1946. p. 27–43.
3. BUTLER, C. H. "Mastery of Certain Mathematical Concepts by Pupils at the Junior High School Level." *The Mathematics Teacher* 25: 117–172; 1932.
4. CROXTON, W. C. "Pupil's Ability to Generalize." *School Science and Mathematics* 41: 627–34; June 1936.
5. DETTMAN, PRISCILLA E., and ISRAEL, HAROLD E. "The Order of Dominance Among Conceptual Capacities." An Experimental Test of Heidbreder's Hypothesis. *Journal of Psychology* 31: 147–60; 1951.
6. DOUGLAS, O. B., and HOLLAND, B. F. *Fundamentals of Educational Psychology.* New York: Macmillan, 1938. p. 318.
7. DREVER, J. I. "The Pre-Insight Period in Learning." *British Journal of Psychology* 25: 197–203; 1934.
8. EWERT, P. H., and LAMBERT, J. F. "The Effects of Verbal Instructions upon the Formation of a Concept." *Journal of General Psychology* 6: 400–413; 1932.
9. FIELDS, P. E. "Studies in Concept Formation." *Comparative Psychology Monographs* IX, No. 2, 1932.
10. GESELL, ARNOLD. *Infant Development: The Embryology of Early Human Behavior.* New York: Harper and Brothers, 1952.
11. HADAMARD, JACQUES. *The Psychology of Invention in the Mathematical Field.* Princeton, N. J.: Princeton University Press, 1949. Chapter VII.
12. HARRIMAN, P. T. *The New Dictionary of Psychology.* New York: Philosophical Library, 1947.
13. HASTINGS, J. T. "Testing Junior High School Mathematics Concepts." *School Review* 39: 766–76; 1941.
14. HEBB, D. O. *The Organization of Behavior.* New York: John Wiley and Sons, 1949.
15. HEIDBREDER, E. "The Attainment of Concepts: I. Terminology and Methodology." *Journal of General Psychology* 35: 173–89; 1946.
16. HEIDBREDER, E. "The Attainment of Concepts: II. The Problem." *Journal of General Psychology* 35: 191–223; 1946.
17. HEIDBREDER, E. "The Attainment of Concepts: III. The Process." *Journal of Psychology* 24: 93–138; 1947.

18. HEIDBREDER, E.; BENSLEY, M.; and IVY, M. "The Attainment of Concepts: IV. Regularities and Levels." *Journal of Psychology* 25: 299–329; 1948.
19. HEIDBREDER, E., and OVERSTREET, P. "The Attainment of Concepts: V. Critical Features and Context." *Journal of Psychology* 26: 45–69; 1948.
20. HEIDBREDER, E., and OVERSTREET, P. "The Attainment of Concepts: VI. Exploratory Experiments of Conceptualization of Perceptual Levels." *Journal of Psychology* 26: 193–216; 1948.
21. HEIDBREDER, E., and OVERSTREET, P. "The Attainment of Concepts: VII. Conceptual Achievements During Card-Sorting." *Journal of Psychology* 27: 3–39; 1949.
22. HEIDBREDER, E. "An Experimental Study of Thinking." *Archives of Psychology*; 1924. No. 73.
23. HEIDBREDER, E. "The Attainment of Concepts—A Psychological Interpretation." *Transactions of the New York Academy of Science.* July 28, 1945.
24. HENDRIX, GERTRUDE. "Prerequisite to Meaning." *The Mathematics Teacher* 43: 334–39; November 1950.
25. HENDRIX, GERTRUDE. "A New Clue To Transfer of Training." *Elementary School Journal.*
26. HOLLINGSWORTH, H. L. *Educational Psychology.* New York: D. Appleton-Century Co., 1933. p. 355.
27. HULL, C. L. "Quantitative Aspects of the Evolution of Concepts." *Psychological Monographs* 28: 1–85; 1920.
28. JUDD, CHARLES HUBBARD. *Psychological Analysis of the Fundamentals of Arithmetic.* Chicago: University of Chicago, 1927.
29. LEVY-BRUHL, LUCIEN. *How Natives Think.* New York: Alfred A. Knopf, 1925.
30. LUECK, WILLIAM R. "An Experiment in Writing Algebraic Equations." *Journal of Educational Research* 42: 132–37; October 1948.
31. MCCONNELL, T. R. "Recent Trends in Learning Theory: Their Applications and the Psychology of Arithmetic." National Council of Teachers of Mathematics. *Arithmetic in General Education.* Sixteenth Yearbook. New York: Bureau of Publications, Teachers College, Columbia University, 1941. p. 268–89.
32. MUNN, NORMAN L. *Psychology.* New York: Houghton-Mifflin Co., 1951.
33. MURPHY, GARDNER. *General Psychology.* New York: Harper and Brothers, 1933.
34. PETERSON, G. M. "An Empirical Study of the Ability To Generalize." *Journal of General Psychology* 6: 90–114; January 1932.
35. PIAGET, JEAN. *Judgment and Reasoning in the Child.* New York: Harcourt, Brace and Co., 1928.
36. REED, H. B. "Factors Influencing the Learning and Retention of

Concepts." *Journal of Experimental Psychology* XXXVI, February–June, 1946. p. 71–87, 166–79, 252–61.
37. REED, H. B. "The Learning and Retention of Concepts. The Influence of the Complexity of the Stimuli." *Journal of Experimental Psychology*. XXXVI, February–June, 1946. p. 71–87, 166–79, 252–61.
38. REED, H. B. "The Learning and Retention of Concepts." II, The Influence of Length of Series, III, "The Origin of Concepts." *Journal of Experimental Psychology*. February–June 1946. p. 71–87, 166–79, 252–61.
39. REED, H. B. "Learning and Retention of Concepts, V, The Influence of For of Presentation." *Journal of Experimental Psychology* 40: 504–11; August 1950.
40. REIDCHARD, S.; SCHNEIDER, M.; and RAPPAPORT, D. "The Development of Concept Formation in Children." *American Journal of Orthopsychiatry* 14: 156–62; 1944.
41. SMOKE, K. L.: "An Objective Study of Concept Formation." *Psychological Monographs*. Vol. 42, No. 4, 1932.
42. SMOKE, K. L. "Negative Instances in Concept Learning." *Journal of Experimental Psychology* 16: 583–88; 1933.
43. SMOKE, K. L. "The Experimental Approach to Concept Learning." *Psychological Review* 42: 274–279; 1935.
44. STEVENS, S. S "The Operational Definition of Psychological Concepts." *Psychology Review* 42: 517–27; 1935.
45. STROUD, J. B. *Psychology in Education*. New York: Longmans, Green and Co., 1946.
46. TATE, M. W. "Operationism, Research and a Science of Education." *Harvard Educational Review* 20: 11–27; June 1950.
47. VAN ENGEN, H. "Analysis of Meaning in Arithmetic." *Elementary School Journal* 49: 321–29; 395–400, February–March, 1949. p. 26–30.
48. VINACKE, W. EDGAR. "The Investigation of Concept Formation." *Psychological Bulletin* 48: 1–31; January 1951.
49. WARREN, HOWARD C. *Dictionary of Psychology*. New York: Houghton-Mifflin Co., 1934.
50. WELCH, LIVINGSTONE. "Behaviorist Explanation of Concept Formation." *Journal of Genetic Psychology* 71: 201–22, December 1947.
51. WENZEL, B. M., and FLURRY, C. "Sequential Order of Concept Attainment." *Journal of Experimental Psychology* 38: 547–57; October 1948.
52. WERNER, HEINZ. *Comparative Psychology of Mental Development*. New York: Harper and Brothers, 1940.
53. WOODWORTH, R. S. *Experimental Psychology*. New York: Henry Holt and Co., 1938.

4. Sensory Learning Applied to Mathematics

HENRY W. SYER

SENSATION, ATTENTION AND PERCEPTION

It seems hardly necessary to justify the importance of sensory learning—it is an essential part of all learning. This is true if one defines the "senses" as any means by which the individual receives stimuli from the environment, and defines "learning" as any adaptation to or acquirement of control over that environment due to training rather than maturation. Sensory learning is concerned with the role which the receptors have in all types of learning, and it is also concerned with any improvement in the techniques of using the senses. These two meanings of "sensory learning" appear in many books; we accept them both. Thus, sensory learning is concerned with the physical aspects of the environment and the physical aspects of the individual which are important for learning.

Sensation is any experience which results from stimulation of the senses; all sensations which are vivid and clear are said to be those to which one is paying attention; and the totality of sensations from a given situation, often from different senses, and organized into a pattern, is a perception. On the basis of perceptions, one abstracts and generalizes to form concepts, carries such concepts in the memory, combines concepts to form higher types of abstractions, and juggles such concepts in mental trial-and-error to form the rudiments of thinking. But we are wandering away from sensory learning merely to show that the senses and their sensations are the basic data of all experience and thought (32: 176–84).

There are eight senses which we shall distinguish: visual, auditory, olfactory, gustatory, cutaneous, static, kinesthetic, and organic (8: 111; 12: 87–95).

The receptor for visual sensations is the eye, and the stimulus is some sort of radiant energy. With the eye we make discriminations of color (hue, brightness, and saturation), distance, and depth (3: 57–101; 33: 78–105).

The receptor for auditory sensations is the ear, and the stimulus is a longitudinal vibration transmitted by the particles of the air. With the ear we make discriminations of sound (pitch, loudness, and complexity), distance, and direction (3: 102–39; 33: 106–17).

The receptor for olfactory sensations is the nose, and the stimulus is a solution of the substance smelled. With the nose we perceive odors formed by combining the four elemental odors: fragrant, acid, burnt and caprylic (8: 96; 3: 146–53; 33: 117).

The receptor for gustatory sensations is the tongue, and the stimulus is a solution of the substance tasted. With the tongue we perceive tastes formed by combining the four elemental tastes: sweet, sour, salty, and bitter (8: 100; 3: 140–46; 33: 122).

The receptor for cutaneous sensations is the skin, and the stimuli are pressures, extreme temperatures, changes in temperature, and electrical and chemical stimuli. With the skin we make discriminations of pressure, pain, cold and warmth (8: 96; 3: 154–72; 33: 125–30, 133–34).

The receptors for static sensations are the semicircular canals of the ear, and the stimulus is the motion of the fluid therein. With this mechanism we make discriminations of balance and equilibrium (3: 176–84; 33: 137–40).

The receptors for kinesthetic sensations lie within the muscles, tendons and joints of the body. With these nerve endings we make discriminations of position and movement of the body (3: 173–76; 33: 134–37).

The receptors for organic sensations lie within the abdominal and thoracic regions of the body and the stimuli which affect them are still not completely understood and analyzed because the nerve endings are so buried in the depths of the body. They lead to perception of such sensations as hunger, thirst, nausea, vascular experiences, respiratory experiences, sexual sensations, and the general "feeling tone" of the body (3: 172–73, 184–85; 33: 131–33).

Among sensations of any one type we can distinguish between different sensations by the four characteristics of quality, intensity, duration, and extension (32: 176–78).

These sensations are the building blocks from which perceptions, concepts, ideas, thoughts and learning are constructed. They are the basis of all learning and find their greatest importance in the fact that any attempt to trace "meaning" back to its origin, or any attempt to clarify "meaning" by explanatory examples and applications eventually leads us to concrete objects and experiences which have involved the senses. True, abstractions may have been built from other abstractions, but somewhere back in the formulation of these ideas is a foundation in sensory perceptions, and often we rush back to these perceptions when the abstractions become vague and confusing.

The busy, buzzing world is presenting us with a flood of stimuli constantly, and we are equipped to make selections from these stimuli and attend only to those which we find interesting or important. Thus some sensations are in the center of our attention and some are on the fringe. It is important to know the factors which control and direct attention (8: 518).

These factors may be in the stimulus or they may be in the observer (8: 523; 32: 182–84; 12: 87–129; 33: 67–76). Factors in the stimulus which are important are those of quality, intensity, duration, and extension—the same characteristics we noted formerly. However, and probably more important, attention is more apt to be caught and controlled by *changes* in these characteristics. Thus certain colors create more attention than others, but any change in color is apt to be compelling. Other changes or differences in proximity, or in contrast with the background material are important. Changes in duration such as sudden appearances or disappearances of the stimulus, and especially motion of the stimulus with respect to the background are attention-getting. And naturally the extension or size of the stimulus is important— the bigger, the more attention-getting.

The factors in the individual which control and direct attention are not so easy to identify. For our purposes we will distinguish three: the novelty of the stimulus in terms of the experience of the individual, the present interests of the individual and the organic state of the individual. Any stimulus which is very familiar, any stimulus which differs from a present, compelling interest of the individual, or any stimulus which is presented when the individual is ill will have less effect than otherwise.

Now that we have discussed the kinds of stimuli which can be presented, and the factors which assure that the stimuli will be attended to, we should discuss the factors which help to organize these stimuli into perceptions. We should discuss the perceptions of space, time, and movement which are needed in learning.

Unfortunately for the scientific approach to education and to psychology, it is impossible to predict the response and reaction to a given set of stimuli. Even in the same person the results may differ at different times due to the internal conditions of that individual, which serve to determine how the stimuli attended to shall be organized into a perception (2: 8-9). Wundt in experimenting with drawings made by children (36: 82) came to the conclusion that perception deals only with the salient, meaningful parts of the object being attended to. This sounds like a truism until one interprets it to mean that stimuli of equal strength need not be remembered or incorporated into a perception to the same degree. These perceptions which belong or fit into a pattern are those which are retained.

Certain factors determine whether perceptions are organized into a pattern: (a) a constancy of form, color, or shape; (b) figure-ground relationships; and (c) the context in which the perception is presented and the experience of the perceiver (12: 125).

This whole subject of perception may have more importance in the teaching of mathematics than in most subjects because there is less inherent meaning in a subject which must stress its abstractness. For example: the perception of the number 164327463 as 16427463 or as 163427463 is not a trivial matter which could be remedied after the stimulus is removed. On the other hand, partial perceptions of words may be sufficient. Would not any of the following be sufficient: "somthing," "someting," "smething," "sometng," "somethnig," "soemthing," or even "smethng," or "smthng?"

What are the chief types of patterns into which perceptions are usually organized in the teaching of mathematics? Certainly the most important are the perception of space, time and movement. Space is most accurately perceived by the eye, time by the ear, and movement by a combination of these two senses (32: 188).

Visual space perception depends upon combinations of the following eight factors: distinctness, shadow, position, relative size, relative motion, sensations of accommodation in the eye, binocular differences in the eye, and sensations of convergence in the eye (8: 65–71). These are used to make discriminations of size, proximity and direction.

Auditory space perception is used to make discriminations of direction and distance. The former are based upon time differences to the two ears, intensity differences, and phase differences; the latter are based upon the intensity and the quality (or timbre) of the sound (8: 91–92).

We are aware of the following types of patterns into which we organize perceptions of time: continuity, succession, length, and rhythm (3: 246–60).

Perception of movement is dependent upon successive stimulation of various nerve-endings in a receptor. However, this stimulation may be caused by a moving object (such as a traveling train moving in front of the eye) or by separate objects stimulating adjacent nerves and causing apparent motion (such as in motion pictures) (3: 260–73).

We have now concluded our survey of the psychology of sensory learning upon which we shall base our discussion of the teaching of mathematics. We have discussed the types of senses, their receptors, and their stimuli; the factors which control and direct attention; the factors which determine how sensations are organized into perceptions; and to some extent the perception of space, time and movement.

RELATIONSHIPS OF SENSORY LEARNING TO OTHER ASPECTS
OF LEARNING

Lest one think that sensory learning is the end and all of learning, or that the piling up of sensations in the brain (much as a phonograph record or a photograph registers its stimuli) is the purpose of improving sensory learning in human beings, it is important to consider the connection between the fundamental sensations and the higher processes of learning. We shall illustrate by using the following higher types of learning: motivation, memorizing, the use of mental concepts, problem-solving, emotional activities, and imagery.

Motivation. Motivation is the condition of the individual which points him toward the practice of a given task and which defines the satisfactory completion of the task (3: 312). The role of sensory learning in motivation is to define the task in terms of the manipulation of concrete objects which may be seen, heard, felt, smelled, tasted, or perceived by some other sense. This is often necessary in order to clarify the purposes of learning and thus lead to motivation. How can an individual point toward a goal if the goal is vague?

For example, if a pupil is asked to discover how many parts of a triangle are needed to determine the triangle, and what combinations of sides and angles will do this, is it sensible to expect him to want to do this if the phrase "determine the triangle" is meaningless? Changing that phrase to "determine the size and shape of the triangle" might make it clearer. However, for many people the attempt to fit sticks together, to see how many essentially different triangles can be made with three given sides, will clarify the idea even more. A little experimentation in drawing triangles with a side of 10 inches, an angle adjacent to that side of 20 degrees, and a side opposite that angle of 6 inches might help to clarify the question even more. By now the pupil may be ready to solve the original problem under his own power because he is motivated to do so. The motivation followed the clarification of the problem and the clarification was the result of interpreting the original problem as a combination of visual, cutaneous, and kinesthetic sensations. Do the sticks fit together? Can a ruler and protractor be manipulated to give one and only one triangle fitting the conditions?

Naturally, the mental triangle, its properties and the conditions which determine it are the important part of this learning, but the physical manipulation made the problem real and thus made the motivation possible. Often, for clarification, a problem will be redefined in simpler terms which are less abstract, but are not sensations to be experienced and manipulated.

Memorizing. We have more real evidence concerning *memorizing* than we have concerning most types of learning (22: 58–66). By

SENSORY LEARNING APPLIED TO MATHEMATICS 105

concentrating upon memorizing one can find correlations between the mental phenomena of learning and the physical phenomena observed in the receptors, nervous system, and effectors. It begins to look as though a physical explanation of learning is possible in terms of conditioned reflexes, synapses and such physiological concepts. Indeed, such correlations are probably more valid the more mechanical and simple the type of learning. Since memorizing is apt to be more mechanical and simple than higher types of learning, we should be able to use the senses to a greater advantage in memorizing than in establishing such advanced types as esthetic and moral learnings.

For example, the learning of addition facts is essentially the memorizing of the results of combinations of objects which are understood, abstracted, verbalized; and then memorized. If one learns to count as the basis for other arithmetical learning then the fact, "five plus three is eight" can be arrived at by interpreting five as ///// and three as ///. Thus five plus three is shown to be /////   /// which then becomes //////// through the meaning of plus. One then can count "one, two, three, four, five, six, seven, eight" or, more probably, "five, six, seven, eight" since the first group is known. The result is then recognized as eight. In order to detach this from the particular sticks, circles, or marks made with a typewriter and generalize to the final meaning of the "addition" fact, it is necessary to present the same abstraction in many forms. Thus, pictures of chickens, apples, chairs, pennies and groups of many other objects are presented to the class so that no particular sensation will be attached to the arithmetical abstraction. Because objects are moved around by the pupils as they learn, we are also presenting groups of cutaneous sensations and kinesthetic sensations as well as visual ones. Auditory sensations could also be used, but groups of sounds or musical notes are so ephemeral that they are not very suitable examples.

An older theory of learning would probably want to explain the use of sensory learning in the memorizing of addition facts as follows: The facts are repeated to the children aloud many times and thus the auditory sensations strengthen the learning of the facts; they are seen on the blackboard, charts, and textbook many times and thus visual sensations strengthen the memorizing; the children say the facts and write them many times and thus more and more sensations, kinesthetic and organic, add to the learning. The results of all this sensory learning is a permanent memorizing of the facts. This is definitely *not* the connection between sensory learning and memorizing which we are presenting here.

Use of Mental Concepts. Sensory learning is necessary to the understanding of concepts at two different levels: (a) when the concept is being developed, and (b) whenever the concept is being applied.

Sensory learning in the development of concepts has been mentioned several times in this discussion already. Here it is again, however: All concepts are abstractions from simpler concepts or from perceptions experienced by the senses. Simpler concepts are those which are nearer to the original sensations from which they were all abstracted. Or, to build in the other direction, one begins with individual sensations which are organized into perceptions and abstracted to form concepts. By adding more sensations and combining concepts we have a hierarchy of concepts which grows further and further from the physical stimuli which are at the base.

One example of concept building is the meaning of "five plus three" previously discussed. Another, quite different, is the concept of a derivative. This concept has fundamentally no more to do with the tangent to a curve than "five plus three" has to do with marks on a piece of paper. The marks are an example of the concept; the tangent to a curve is a particular use of a derivative. Nevertheless, the derivative is usually introduced by discussing tangents to curves because it is more concrete. This is the same thing as saying that it is nearer to physical stimuli. Which of the following two statements is more apt to suggest visual and kinesthetic sensations? (a) The tangent to a curve is the limit of a secant through the point of tangency and another point on the curve as the second point approaches the first as a limit, or (b) the derivative of a function is the limit of the ratio of the increment in the value of the function (which corresponds to an increment in the independent variable) to the increment in the independent variable as the latter increment approaches zero. Is it any wonder that the first approach is used as an introduction? The same reasoning is used in good teaching to introduce the definite integral by means of area, the concept of mathematical induction by talking about knocking over a row of dominos, the concept of subtraction by the phrase "take away," the concept of percentage by the picture of a pan of fudge ten squares on a side, the concept of a coordinate system by locating positions on a map, and the concept of fractions by circles or pie plates cut into sectors. All these show the use of sensory learning to build concepts.

An "application" of a concept which has been well understood and established is, by my definition, a step from that concept in the direction of perceptions. If it is a simple concept, the application may take us back to such perceptions in one step; if a more complicated concept, merely in that general direction.

For example, the concept of measurement as the comparison of an unknown quantity with a standard called a unit may have been built by measuring desks, blackboards, pieces of paper, children's heights and flower pots in the schoolroom. Then the class could be asked how one could measure easily the distance from the school to the firehouse down the street. Someone might come up with the idea that the yardstick (their largest unit) could be used to measure a piece of string (which was considerably longer than the yardstick) and then the piece of string used to measure the distance to the firehouse. The concept had been understood, and the application consisted of a description of the physical steps that one would go through in creating a new, larger, and more convenient unit. Indeed, if you try this with a class, you may well find, as the writer did, that they put in many details of how to hold the string, how to tally by marks on the blackboard the number of yards found in it, and other descriptions of the sensations which they expect to experience in applying the concept of measurement to the fire house problem.

Problem-Solving. One use of sensory learning in problem-solving is the translation of a problem into some sensory analogue or model, with manipulation of that model, either physically or mentally to solve the problem. This process is similar to the method of analytic geometry—a problem in geometry is translated into algebra, the algebra is manipulated, and then the result is translated back into geometry as the answer to the problem. This is similar, but not the same. Algebra is more abstract than geometry, farther from perceptions, so the method of analytic geometry is actually the reverse of the method we are describing. Analytic geometry is powerful, not because the algebra is easier to visualize, but because it is more formal, and therefore its rules of manipulation are more completely understood and more thoroughly organized.

An example of this important use of sensory learning is the language of n-dimensional geometry. The interior of a sphere in ten-dimensional space has nothing whatsoever to do with our perceptions of the space in which we live. It is merely a mathematical shorthand which appeals to our perception of three-dimensional space, and, by analogy, gives a method of remembering a certain algebraic inequality concerning ten variables. Sensory learning is used to supply a framework for our thinking about abstractions which have nothing to do with our senses at all. Physicists use this method all the time in constructing models of the atom. Another example is their insistence on a wave theory of light or a corpuscular theory or some combination of these theories which can have an interpretation in terms of our sensations. Recently, mathematical expressions have been said to hold the only claim to "reality," but the less esoteric of us still yearn for the physical models as explanations.

Cole and Bruce have some quotations which are particularly apt here (6: 517–18):

> The building of the model, the drawing of the diagram—either in imagination or with physical materials—is a crucial step in the human thought process. The architect thinks with his drawing board and instruments, the machine builder with his model.... It is this ability to build a thought model which transforms the direct, bungling, blundering behavior of the unthinking child into the thoughtful, planful, reasoning behavior of the adult.... What a thought model does, in addition to giving us something to manipulate in our planning, is to provide a host of suggestions of possible manipulations...: Science is ...a system of models, symbols, relationships; and, once we have

mastered its network of relationships, we turn to it quickly when we are confronted with a problem.

A more elementary example from mathematics is shown in Figure 1 which gives a graphical method for deriving the formula

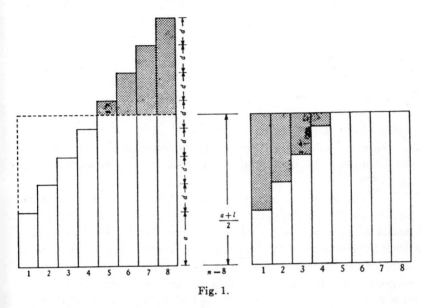

Fig. 1.

for the nth term of an arithmetic series,

$$l = a + (n - 1)d$$

and also for deriving the sum of the first n terms,

$$S = n\left(\frac{a + l}{2}\right)$$

Another example is seen in Figure 2, which is based upon an Indian method of proving that $a^2 + b^2 = c^2$. In both these examples the pictures are analogues of certain algebraic manipulations, and make them easier to justify and to remember.

Thus the greatest contribution of sensory learning to problem-solving is the construction of models which can be manipulated to produce sensations, actual or imaginary, to aid in finding solutions to the problems.

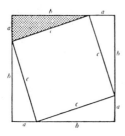 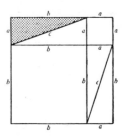

Fig. 2.

Emotional Activities. Next, we should like to consider the connections between sensory learning and emotional activities. We shall consider the development of aesthetic appreciation and of attitudes.

Aesthetic appreciation certainly calls first to mind the appreciation of paintings, sculpture, engravings, architecture, music, the works of nature, good cooking, perfume and other sensations. Even the elements of appreciation, the factors which are discussed in order to reach a decision of aesthetic value, are reminiscent of physical stimuli: balance, rhythm, emphasis, movement, unity, and contrast. These terms are applied to nonrepresentative art as well as representative. As soon as the objective is to represent photographic fidelity in the graphic arts, imitation in music and perfume, and the semblance of something which it is not in cooking, then all the senses are called upon to make comparisons with the object being represented. This shows that whether art is intended to be lifelike or "pure" or whether it is judged by one standard or the other, the judgments are always based upon concepts which have their foundations in sensory learning.

The examples for aesthetic appreciation which can be drawn from the teaching of mathematics are few; not because there is no connection, but because little use of them is made in teaching. The balance of geometric forms is a standard cliché in relating mathematics to art. What teacher has explored the possibility of teaching rhythm and its appreciation in mathematics classes? Certainly a consideration of various rhythms on the drums and the mathematical relationships between the duration of the sound and the duration of the silence, the pattern of recurrence of these sounds and silences and the recognition of waltz, rumba and marching rhythms by the mathematical characteristics of their graphical representations is an untried field. It is not destroying the enjoyment of art to understand it. Analysis leads to appreciation.

The second of our emotional activities to be considered, "development of attitudes," is taught to a large extent only indirectly in mathematics. Mathematics, in common with all other subjects, is responsible for the social, personal, moral, and religious attitudes of students. Since no course is entitled "Attitudes," nor should it be, this phase of teaching is shared by all. It is a perfectly fair question whether we are carrying our share of teaching these general attitudes. Sensory learning plays its part this way— incorrect attitudes are often the results of incorrect understandings which, in turn, are based upon incorrect sensory perceptions. Attitudes can be derived from experiences with mathematics. If it is presented at too abstract a level it will result in frustration and thus antagonism to mathematics, the authority of the school, and possibly to people who can do mathematics well. Therefore, in some cases, a more concrete approach could encourage correct attitudes.

There are attitudes which are peculiar to mathematics and which have a definite basis in sensory learning. The best example to use here is summed up by the word "numerology." In its strictest pseudo-scientific sense this subject juggles numbers derived from a person's name, birthdate, the present date, his telephone number and other sources and finally arrives at clear-cut decisions concerning his vocation, marriage, financial investments and anything else on his mind. We see here the groping of people for a certainty outside themselves in situations where there can be no certainty, a place to transfer the blame for a decision. Do we not have a responsibility to disassociate such duties from mathematics? That subject has enough to do with its legitimate tasks.

Superstitions using mathematics are common. Their interest for us lies in the fact that many who believe them say they have reason for doing so. They have known of three deaths coming together many times; when they carry four-leaf clovers everything turns out much better than when they do not; 13 is unlucky because there were 13 at the Last Supper; 13 is unlucky because there are 13 steps to a gallows and 13 ridges on a hangman's knot. Here is a definite appeal to sensory perceptions to justify the superstitious attitudes. The teacher's task is to turn this same trust in sensory learning into a scientific experiment to test some superstitions. Not only would faith in irrational mathematics be attacked, but an illustration of the scientific method would be displayed.

Imagery. Our final connection between sensory learning and higher processes in learning will concern imagery. We can classify images into six types: after-images, eidetic images, memory images, imagination images, dreams, and hypnagogic images. Thus imagination will fall into place at just one of the types of image (3: 360–68). Imagery is usually discussed in terms of visual imagery, but it can result from any type of sensation.

In general, an image is the reproduction of a past perception, in whole or in part, in the absence of the original stimulus. The lingering of the perception for a few seconds or minutes after the stimulus is removed is called an after-image. Eidetic images are remarkably clear images which persist after the stimulus is removed which carry details apparently not originally attended to, and are sometimes under the control of the individual. Memory images are the most familiar type and are less clear than after-images or eidetic images but can be recalled after longer periods of time. Imagination images combine parts of memory images to create images which never were experienced in that form as the result of an external stimulus. Dreams are images experienced while we are asleep, and their origin in past perceptions is vague and not fully understood. Hypnagogic images are similar to imagination images but occur in that half-and-half state between waking and sleeping. For the teaching of mathematics the most important types of image are the memory image and the imagination image.

Memory images are economy measures. It is certainly more efficient to be able to call up the perception caused by some past

SENSORY LEARNING APPLIED TO MATHEMATICS 113

stimuli. Without this ability there would be no thinking or learning, only experiencing. The ability to control memory images varies with many factors both within and without the individual. We cannot amplify this subject of memory without losing the thread of our present discussion.

Since learning is concerned with the retaining of correct memory images (among other things) and thinking is concerned with the manipulation of mental images (among which are memory images based upon sensory perception), it is easy to see the importance of sensory learning for memory images. The success of learning and thinking depends in part upon the ease of recalling such images and the clarity of such images when they are recalled. This ease and clarity will depend upon the conditions under which the original perception took place. If the perception has been well organized, both within itself and also into the experience of the individual, the memory image will be easy to recall and clear.

Let us apply this to the memory of the rule for adding signed numbers in beginning algebra. In order to organize the perception well within itself we need to show many examples, using both small numbers and large; with both numbers plus, both minus; with the one of larger absolute value plus, the one of larger absolute value minus; with the plus number above the minus one, the plus one below, the plus one to the right, the plus one to the left; with possibly fractional and decimal forms; and finally with signed algebraic expressions. From all of these the rule we wish remembered will emerge in a memory image that should be strong and clear. In order to organize the perception into the experience of the individual we built the concept from concepts already known and apply it to other concepts already known. In the example of the signed numbers we use such ideas as thermometers, assets and liabilities, above and below sea level, or any other concepts already a part of the pupil's experience. Thus, a well-organized perception leads to a well-established memory image.

The other type of image, the imagination image, useful in learning and thinking, may be even more important. In order to arrive at new conclusions, not identical with perceptions previously experienced, a pupil must have the ability to break loose parts of memory images and combine them into new, imagination images. This type of activity is a necessary prerequisite to all physical creative activity whether it be practical, such as the design of a new type of aircraft, or artistic, such as some new, nonobjective painting. The concept must appear in an imagination image before it can appear as fusilage or oil paint. If practical and artistic progress is made by those with active imaginations, what can mathematics do to encourage such imagination?

Imagination is needed in mathematics to suggest possibilities for proofs of originals in geometry; to suggest combinations for factoring a quadratic polynomial; to suggest that variable which should be "x" in a word problem; to suggest things to buy from the playstore in arithmetic; to suggest the best way to place the co-ordinate axes at the beginning of a problem in analytic geometry. What role does sensory learning have in each of these situations? In solving geometric originals it is useful to have some flexible devices to illustrate possible arrangements of geometric parts in the figure. Such devices made of cardboard, wooden strips, and elastic, as the Burns Boards (23: 379–95) give sensations which can be changed at a speed nearer that with which the mind changes from one image to another than do less flexible methods of illustration. Of course, this is a crutch to the imagination, but such crutches should be used until the imagination is strong enough to walk without them. It is up to the teacher to withdraw the use of the crutch as fast as the pupil's abilities allow. An extended use of such visual aids is just as weakening to the imagination as the refusal to use them at all is stifling.

In factoring quadratics it may be well to have cards with spaces left blank and other cards with numbers on them to fill in these spaces in order to show, visually, the possible combinations which should be considered as factors. Even a vocal listing of such possibilities may encourage the pupil's imagination to make similar exploratory lists later for himself. In setting up a word problem, a blackboard demonstration of possible choices for "x" may be long and unwieldy, but it lays the framework for the pupil's imagination to make mental listings and manipulations for himself. At an elementary level we want children to invent their own problems in addition by playing store and making lists of items to buy. Before they do this in their imagination it is helpful to have real objects, tin cans, boxes, and paper bags, marked with prices, for the children to manipulate and add. Later the imagination images of new combinations of these prices create addition problems for the pupils. In teaching problems in analytic geometry it is well to solve the same problem several times with the axes in different positions and to see what effect this has on the simplicity and neatness of the procedure and the solution. Later the students can make a choice of the position of the axes with their imagination suggesting the various results. Without the experience of the actual sensory perception of the results at one time, one's imagination has nothing upon which to base its creativeness. These examples show the need for sensory learning to stimulate the imagination in mathematics.

There are other reasons why mathematics teachers should consider the role of imagination in their teaching. Some people turn very abstract ideas into very concrete forms to aid the memory and use of these ideas. Some of these forms are very artificial. For example, Galton (11) made a study of the number forms which some people create to visualize the relationships between numbers. Some people think of the number 1 at their left elbow, and then the numbers 2 through 10 in a sweeping curve out in front of them, other numbers trail off into space in a well-defined curve which stays in their imagination in that same form to help them with their quantitative thinking. Teachers need not encourage such imagination images, but they should realize that they exist in the imagination of some people.

Our last appeal for an understanding of imagination images by the mathematics teacher is based upon the feeling that creativeness and expressional activity can be a part of all school subjects, and that mathematics is one which is eminently suited to such activity. Instead of making our teaching a succession of memory images, we should search for every opportunity to encourage the free imagination, the speculative twist of perceptions, the courage to think and talk about what would happen *if*, the shared delight of the new idea, the explorer's enthusiasm for new physical territories transferred to the realms of the mind, and the feeling that mathematics is also what you make it, not only what it has been made.

We started with the mundane facts of the physical stimuli around us and have ended with flights of the imagination. Sensory learning becomes a part of all mental activity and we have tried to trace its influence in motivation, memorizing, the understanding of concepts, problem-solving, emotional activities, and in imagery.

RESPONSIBILITIES WITHIN MATHEMATICS TO IMPROVE THE USE OF THE SENSES

We have now given the psychological basis for sensory learning and tried to show why this type of learning is needed for all types of learning. It is the chief purpose of our entire discussion to improve the teaching of mathematics by improving the use of sensory learning in that subject, but we shall stop our progress for a moment to point out the meaning and importance of the converse problem—how can the teaching of mathematics contribute to sensory learning in general?

Schools have a responsibility to improve citizenship, health habits, and the use of the senses to gather knowledge through perception. There are other responsibilities also, but these will serve to illustrate the types of objective which pervade all subjects. Of course mathematics teachers do not have time to teach the three topics listed above, but neither do any other teachers! We may get teachers to agree that the major responsibility for each topic lies in one area—citizenship in the social studies, health habits in physical education, and sensory learning in art. But does not science use sensory learning in its laboratory experiments, English in descriptive writing, French in imitative acquirement of pronunciation, commercial education in typewriting, social studies in map reading, and physical education in tumbling? Where is the opportunity in mathematics to improve the use of the senses?

In order to give definite examples within mathematics we shall show how the teaching of mathematics can help to develop the motor skills, to improve attention, to improve perceptual discrimination and estimation, and to eliminate perceptual illusions. In order to make the examples from the teaching of mathematics clear and distinct, from now on they will be numbered in order.

SENSORY LEARNING APPLIED TO MATHEMATICS 117

Development of Motor Skills

Some writers prefer to call "motor learning" by the title "sensori-motor learning" or even "perceptual-motor learning" (19: 144-49). The teacher who wishes to teach motor skills should know methods of performing the skill and specific practice techniques both of which have proved successful in the past. In explaining such techniques it will be found useful to use sketches, diagrams, slow-motion motion pictures, skilled performances, specially designed drills, and carefully formulated verbal directions. There

Fig. 3.

are many informal occasions in the teaching of mathematics when motor skills are used and developed.

Example 1: Some of the elementary formulas for volumes can be illustrated by constructing models of the volumes and comparing the amount of salt, sand or dried beans which fill the models. The pupils in Figure 3 are pouring dried split peas from a pyramid to a prism with the same base and altitude to show that the pyramid has one-third the volume of the prism. The boy at the left is holding a model which is made of a cone and cylinder with the same base and altitude and shows the same relationship. These models are made of plastic, but cardboard would do very well.

Example 2: The construction of computing devices, such as

those shown in Figure 4, and their use will give practice in motor skills. The Napier's rods at the right are simple to make and are understandable to any pupil who understands the multiplication tables. They are used to avoid the memorizing of such tables, but are inconvenient to carry around. Simple, flat cardboard rods can be made rather than the wooden ones illustrated here. The familiar devices shown with these rods are examples of abacuses both commercial and pupil-made, and of a modern computing machine. There is a very beneficial, fine motor skill involved in

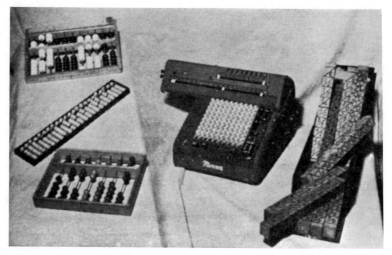

Fig. 4.

the manipulation of all these instruments. Naturally, a great deal of mathematics is also learned.

The whole field of constructing and manipulating multisensory devices is applicable to the development of motor-skills. Moreover, this is better than buying such aids. Since their design can be fitted to the needs of the class, a great deal of mathematical learning is involved *while* they are being made.. They give the pupil a real sense of contribution, their construction develops a sense of independence which our usual dependence upon mechanical devices inhibits, and homemade devices are cheaper.

Some people go so far as to say that all counting, even silent counting, is based upon motor activities and discrete nerve im-

SENSORY LEARNING APPLIED TO MATHEMATICS 119

pulses which place an upper limit on the speed which can be obtained (14: 285). Thus oral counting and silent counting do not differ by much in their speeds.

Improvement of Attention

If we are to improve the ability of an individual to attend to stimuli we must look for methods to improve the factors which control and direct attention and which lie within the observer. Attention will improve if situations are encountered which call for attention and which appeal to the interests, attitudes, set, and experiences of the observer.

Example 3: Here is a teacher demonstration which should appeal to the interest and attitudes of most pupils at the eleventh- or twelfth-grade level. The teacher has in the mathematics classroom an optical bench with light source, screen and various lenses in holders. Object-distances and image-distances for each lens are tabulated, and enough of them for each lens to allow conclusions to be drawn. The pupils are then turned loose to find some constant relationship which these distances satisfy. With or without help they should come to the conclusion that the sum of the reciprocals of the object-distance and the image-distance is a constant for each lens. Since the constant is larger as the lens is thicker we will take the reciprocal of this constant and call it the focal length. So our experiment is summed up by the equation:

$$\frac{1}{f} = \frac{1}{d_o} + \frac{1}{d_i}$$

Compare this method of approach with a straightforward one beginning with the equation in x, y and z and ending with a simple statement that lenses are an illustration of this general type of functionality. Which approach would command the more attention?

Example 4: We expect pupils to pay attention to geometric forms in art around them; why not give them some sensory experiences which will improve their set and thus facilitate such attention. The construction of window transparencies (31: 86–87) by the pupils will give them that background, and will improve their attention to such forms.

Our contention here is that the improvement of attention in many separate situations by many different school subjects is the only way to improve attention in general. It is impossible to design exercises to improve attention without having attention paid to something specific.

Improvement of Perceptual Discrimination and Estimation

The improvement of estimation is essentially a responsibility of mathematics teachers and one that has been greatly neglected because nothing very definite has been suggested to them.

All measurement is really a comparison of an unknown amount of material with a standard called a unit. Estimates of physical quantities are really rough measurements when a low degree of accuracy is sufficient.

There are four ways in which estimating is used: (a) estimating results of arithmetical operations, (b) estimating values by using judgments based upon past experience, (c) naming the number of units which characterize a given amount of physical material, and (d) choosing an amount of physical material which is equal to a given number of units. The last two are those which we call "estimates of physical quantities" and are the ones which can be improved through sensory learning in mathematics.

It is impossible to form correct concepts of the meanings of physical units without some concrete experiences with objects measured in these units. Moreover, in order to assure some proficiency is estimating physical quantities, specific practice in these skills must be given in school. Pictures of the material needed for Examples No. 5–8 will be found in Figure 5.

Example 5: One of the basic decisions used in estimating is that of "greater-less." We can devise an exercise for the kinesthetic sense by taking a series of boxes (eight is a good number) about 1" x 3" x 4", and all exactly alike, and weighting them with lead dress weights. Put no weight in the first, one in the second, two in the third, and so on. Fasten the weights together and to the box so they will not rattle. Fasten the boxes shut with scotch tape, and paint the numbers 1 through 8 on the bottoms of the boxes out of sight. Shuffle them on the table and ask someone to arrange them in order of weight. When he has finished,

SENSORY LEARNING APPLIED TO MATHEMATICS 121

turn the boxes over to check the order of arrangement. Discuss with him the method he used to make his decisions, since this verbalizing helps to analyze the method. By leaving out every

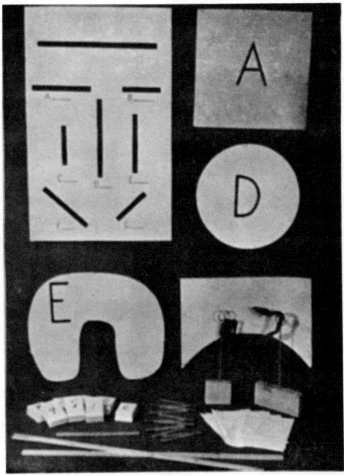

Fig. 5.

other box, or using just the first and last, various levels of difficulty can be constructed to fit the situation.

Example 6: Another basic decision in estimating is "how many times." The teacher and the pupils should all have yardsticks in front of them. Without announcing the distance, the teacher holds her hands a foot apart (measuring it on the yardstick) and tells the class to hold their hands twice as far apart. Then they measure the distance on their yardsticks and tell the teacher what it is. Various distances and amounts should be practiced. Simpler exercises such as drawing lines three times as long as those on a mimeographed piece of paper, give practice in estimating distances closer at hand and show new techniques of estimating.

Example 7: Flags on sticks in wooden blocks should be prepared for an exercise in estimating lengths. The pupil can be

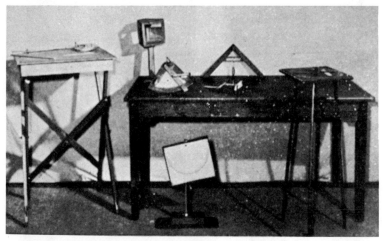

Fig. 6.

asked to place two flags 20 feet apart and the result measured with a steel tape. Then another pupil could be asked to move the flags to a new position and a third pupil to estimate how far apart they are. These two exercises illustrate the two types of estimating of physical quantities. Each exercise should be followed by enough discussion to assure that the class knows ways to improve skill in estimating.

Example 8: The estimation of shape is a complex skill. Visual estimation can be improved by having a page of shapes mimeographed, only one of which matches that of a larger shape held up before the class. This exercise can be made easy or difficult. The discussions which follow are very interesting and reveal novel ways of estimating shape. Blocks cut out of plywood and nailed to a plank provide an interesting blindfold exercise. Two of the shapes on each plank should be the same. It is easier if they are in the same relative position, but they need not be. The pupil is to find the similar shapes using the cutaneous sense without visual help.

Example 9: In Figure 6 will be seen a number of measuring devices which have scales which need to be read and which require some perceptual discrimination. The rulers, tapes, micrometer caliper, and surveying instruments can all be understood by secondary-school pupils and can be used to measure distances and angles both inside and outside of the classroom. The discussions of the construction of scales, the reading of scales and the ways to improve accuracy in readings are very beneficial.

Example 10: Field exercises (28) with the surveying instruments just mentioned can be used to improve the perception of auditory stimuli by planning for complex commands to be transmitted by pupils to each other.

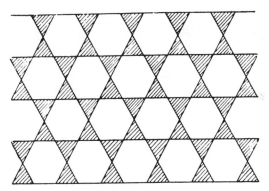

Fig. 7.

Example 11: A fascinating exercise is that of finding how many ways one can cover a plane using congruent regular polygons (29: 59–67; 18: 273–82). If they must all have the same number of sides there are only a few solutions, but if several types may be used (the only restriction being that each type have congruent regular polygons of the same number of sides), then many patterns can be made. Figures 7 through 12 illustrate some of the results. The

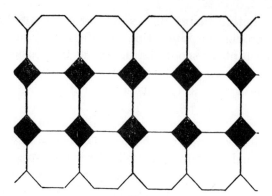

Fig. 8.

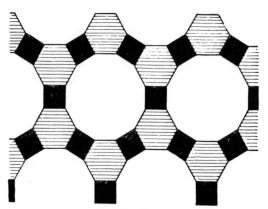

Fig. 9.

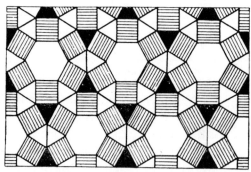

Fig. 10.

SENSORY LEARNING APPLIED TO MATHEMATICS

class may wish to make puzzles from some of their designs to submit to other members of the class.

These are but a few of the types of exercise which teachers can devise to improve the perceptual discrimination and ability of

Fig. 11.

Fig. 12.

their pupils to estimate. Such exercises within the mathematics class should improve these skills in other subjects as well.

Elimination of Perceptual Illusions

The improvement of motor skills, attention, and estimation relate to changes which can be made in the individual. There are some arrangements of stimuli which lead so consistently to the wrong perception that the trouble seems to lie outside the individual. These are called illusions. We know such perceptions are wrong because instruments with more accuracy than our senses convince us of our error. The occurrence of these illusions cannot be prevented, but we can convince our pupils that their sensory learnings sometimes need to be checked against more objective means of observation. The usual illusions of parallel lines and unequal distances will not be given here since they are so familiar.

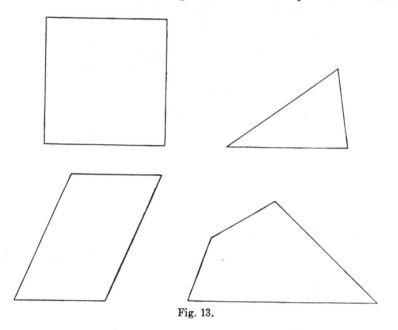

Fig. 13.

Example 12: The center of gravity of a figure may be perfectly understood but our senses place it in the wrong place (5: 136; 30: 25). Try finding the center of gravity of the drawings in Figure 13 using your eyes alone. The geometric constructions for them are shown in Figure 14. We have used the notation $M(A, B)$ to mean the midpoint of the line segment AB. X in each case is the center of gravity. In using these in class it is helpful to have pieces of cardboard in the shapes shown, and to check the geometric constructions for center of gravity by balancing these cardboards on a sharp point or by hanging them from two different points on their edge and using a plumb bob to find two vertical lines which will intersect at the center of gravity.

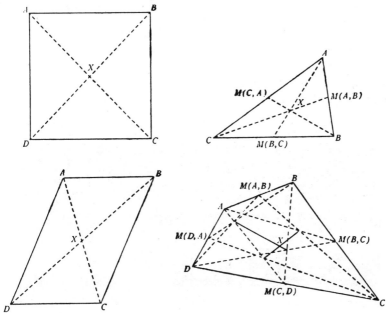

Fig. 14.

Example 13: There should be wooden models similar to the geometric shapes in Figure 15 and thick enough to roll along on edge. Pupils should be asked whether any other shape except a

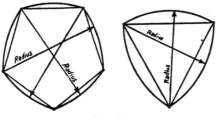

Fig. 15.

circle will roll along a line so that the highest point is always the same distance from the line. Their mental imagery usually says no. Even when the shapes are shown to them they doubt it.

Actual experiment is necessary to convince them. The theory by which these figures are constructed might be given to them, or they could be asked to work it out for themselves (24: 178–79).

Example 14: A common misunderstanding about conic sections is the difference between the shape of a parabola and of an hyperbola. By showing how these are cut from a cone made of clay, or from separate, congruent cones made of wood (such as those in Figure 16), we are able to dispel the illusion that "the parabola and hyperbola are really the same curve, but in one case you have two of them." The best way is to have the pupils cut the

Fig. 16.

clay or the wood themselves. The next best way is to let them watch such a procedure and handle the results.

By using sensory learning we can eliminate some perceptual illusions and reduce the possibility of others. Once a person is convinced that his senses are not to be trusted in all cases where fine degrees of accuracy are needed he will beware falling into such perceptual traps in the future. A rational use of the senses is sometimes necessary.

These 14 examples have tried to show teachers of mathematics how sensory learning in their subject can have general effects improving the uses of the senses as well as specific uses in the learning of mathematics. We have found examples from the development of motor skills, the improvement of attention, the improvement of perceptual discrimination, and the elimination of perceptual illusions.

IMPROVING THE TEACHING OF MATHEMATICS THROUGH BETTER USE OF SENSORY LEARNING

It seems to be about time to get to our real reason for the present discussion—the improvement of the teaching of mathematics. On the other hand, the three sections preceding this one should certainly have some effect on mathematics teaching. The first section summarized the background in psychology upon which the later discussion is based, the second section showed how more complex types of learning depend upon sensory learning, the third section considered the inverse problem—how mathematics can improve the use of the senses, and the present section illustrates the preceding discussion with examples from the teaching of mathematics.

We shall not attempt to introduce any new psychological ideas, nor to present any new mathematics, merely to bring the fields of psychology and mathematics closer together by very definite teaching suggestions. In fact, this whole section will be nothing more than one example after another arranged in the same sequence as the ideas in parts I and II.

Improving the Teaching of Mathematics Through Visual Sensations

Example 15: We have already noted that geometry is more concrete than algebra. Thus, when possible, algebra should be interpreted in geometric terms to clarify the concepts. In factoring, each pupil could be asked to mark a piece of paper or cardboard as in Figure 17a and cut it out with scissors to see that

$$(a + b)^2 = a^2 + 2ab + b^2$$

In the light of our present discussion it is important that (a) each pupil does this for himself, (b) actual cutting be done with scissors since it is much more effective than merely fitting together puzzles which have been previously cut out, (c) any demonstration by the teacher be presented after, not before, the drawing and cutting by the pupil. The other illustration, Figure 17b, shows a similar dissection of a wooden cube to illustrate that

$$(a + b)^3 = a^3 + 3a^2b + 3ab^2 + b^3$$

It is probably too much to expect that this model be constructed by each pupil; instead, he will merely watch it being demonstrated and then manipulate it himself. By using these devices properly we can also show the expansions of $(a - b)^2$ and $(a - b)^3$.

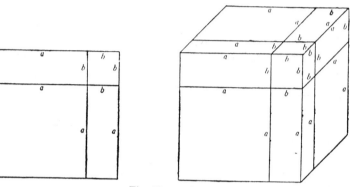

Fig. 17 a and b.

Any verbal description of how these pieces should be manipulated is very long and involved compared with the experience of trying to manipulate them, and that is just the essence of our argument—the sensory learning is more direct and easier to follow than the verbal generalization.

Example 16: The three diagrams in Figure 18 are geometrical representations of the atom. In what sense are they mathematical realities and how are they attempting to use visual sensations? Certainly no scientist would proclaim that we might some day design a microscope strong enough so that an atom can be perceived by the eye in the form of one of these pictures. They are visual aids to thinking. Some of the properties which we have discovered in the atom are remembered and manipulated mentally easier by associating them with the particular visual sensations from these diagrams. It seems appropriate for us to use these as examples here because so much mathematics we try to teach is highly abstract and needs visualization (or other sensory association) similar to that shown here for the atom. Indeed, let us not be too quick to draw the line between mathematics and its applications lest we cut deeply into our subject and, fearing to teach some applications, exclude some mathematics. It may be well

SENSORY LEARNING APPLIED TO MATHEMATICS 131

worth the time in mathematics to draw and construct diagrams for atoms, discussing why cubes, circles and ellipses have been chosen to represent physical concepts.

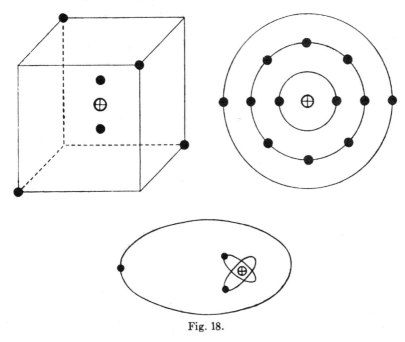

Fig. 18.

Improving the Teaching of Mathematics Through Auditory Sensations

Example 17: A very effective demonstration concerning area and volume can be presented using gunpowder and clay. A cube of clay two inches on an edge is constructed and its area of 24 square inches is computed. As it is cut into halves, quarters and eighths the area is shown to become 32, 40 and 48 square inches, while the volume stays the same. This demonstrates the idea that the total exterior area of a constant volume increases as the size of each piece decreases. In order to show a practical application, we ignite equal weights of two types of gunpowder, and note that the time to burn, brilliance, volume of sound, and density of smoke are functions of the size of particles of the gunpowder. The class never forgets this demonstration.

Example 18: Oral arithmetic drill is a venerable method of varying practice by using the auditory sense. It is not to be spurned because one is seldom asked to do such arithmetic problems after merely hearing them. It is another method of presentation; and the facts to be learned are strengthened because they are abstracted from many types of situations, from many avenues of sensation. Commercial records of arithmetic drill have been made available, but there is still a tremendous amount of investigation open to determine the types of aural presentation which are interesting and effective. Are there recorded stories which should be available to schools as well as the books and sets of pictures we now have to introduce number concepts? What background in the history and present-day applications of mathematics should become available on recordings so that scholarly information and professional radio performance might increase their value?

Improving the Teaching of Mathematics Through Olfactory and Gustatory Sensations

Example 19: A glass of water at room temperature is placed before a blindfolded student. Another student adds a gram of sugar or salt once in a while and the blindfolded student tastes it every so often. It should be tasted more often than material is added. The number of grams when the taste is first detected should be recorded. A distribution for each student or for a whole class will lead to discussions of statistical concepts. If some students are found who are fairly consistent in their decisions it might be well to continue the experiment with different amount of water to begin with and to see whether the amount of material and the amount of water seem to be linear functions of each other. Another series of experiments can be devised as the time which a given substance (ammonia, vinegar, or oil of citronella) is detected at a given distance.

Improving the Teaching of Mathematics Through Cutaneous, Kinesthetic, and Organic Sensations

Example 20: Previous discussions have referred to the use of these sensations. A great deal of fun and learning of mathematics

SENSORY LEARNING APPLIED TO MATHEMATICS 133

will result from experiments with sensations to illustrate the Weber-Fechner Law in psychology (27: 91–103). In one form this law says, "Equal differences between sensations means equal proportion between stimuli." A weight of 40 grams is increased gradually until 75 per cent of the time it is judged to be heavier than the original. We shall call this the threshold of discrimination for 40 grams. The same experiment is repeated for 80 grams and for 200 grams. The graph of the results will lead to the topic of exponential variation and logarithms. Many other experiments are possible.

Example 21: It is known that some people find it necessary to count to establish their addition facts for some time after they

Fig. 19.

are understood. In order to facilitate this, they actually locate parts of the figure which they touch with a pencil or with their finger and count. After the physical movement of the hand is unnecessary, the eye will follow that pattern as an aid. One student reported the patterns shown in Figure 19.

Example 22: The practice of using one's fingers to count or to keep track of the figure to carry in addition is fairly common. It seems that such practices should not be condemned, in some cases they should even be taught as a first sensory approach. The danger is that such immature solutions to problems will persist. Thus pupils should not be made ashamed of having counted on their fingers or made to feel that they have to hide it; it is a natural approach to numbers through sensations. They should be shown more efficient ways of reaching number conclusions, but it may be possible that some people will always count on their fingers.

Example 23: The entire subject of models to be used in the teaching of mathematics is connected with the need for sensory learning. Geometric forms may acquire considerably more meaning if they are handled in concrete form. The models in Figure 20

Fig. 20.

show some regular polyhedrons on the right and some plaster models of spheres on the left. The regular polyhedrons can lead to fascinating discussions of relationships between the number of edges, vertices, and faces. Euler's theorem can be arrived at inductively with a little help and checked for non-regular polyhedrons. Pupils can be asked what figure results when the center of each face of a polyhedron is connected to the centers of all faces adjacent to the first one. This is a little difficult to visualize without the help of a model. If the model can be touched and turned about, the learning is that much easier. The concept of reciprocal figure can be built as shown by the following table:

ORIGINAL POLYHEDRON	VERTICES	EDGES	FACES	INSCRIBED POLYHEDRON
Tetrahedron	4	6	4	Tetrahedron
Hexahedron	8	12	6	Octahedron
Octahedron	6	12	8	Hexahedron
Dodecahedron	20	30	12	Icosahedron
Icosahedron	12	30	20	Dodecahedron

After the table is built by laborious counting, the power of methodical thinking can be emphasized by deriving some of the more difficult facts from some of the simpler. For example, the twelve faces of a dodecahedron can be counted fairly easily; then the number of edges can be found by saying, "Twelve faces each with five edges gives 60 edges, but each edge of the solid figure serves two faces, therefore the solid has 30 edges."

The spherical figures in our picture are made of plaster and show some of the difficult concepts such as spherical polygon and spherical triangle, lune, zone, spherical pyramid, spherical wedge, spherical segment, and spherical sector. On the whole spheres, painted with white enamel, one can draw spherical triangles, polar triangles, quadrants, and erase them. The manipulating of these figures is just as important as seeing them.

Improving the Teaching of Mathematics Through the Use of Muscular Skills

Example 24: The concept of locus can be developed in the following manner. Either a bar or horseshoe magnet should be placed in the center of a fairly large piece of white paper, about 20" x 30". A very small compass is then used to trace the lines of force from one pole to the other. It is stepped along from point to point until the locus develops. This is a fascinating occupation and once a diagram is started it can be left for many pupils to work on and to build as complete a pattern of lines as time allows. Figure 21 shows a typical result for a bar magnet.

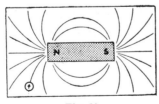

Fig. 21.

Example 25: Another use of muscular skills is the creation of interesting designs by curve-stitching (21: 82–85). There are three basic mathematical ideas which are used in these designs, and they are shown in Figure 22. The first is a regular polygon

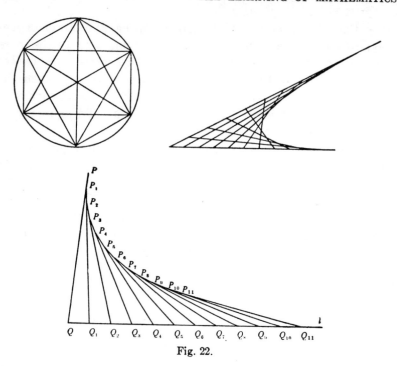

Fig. 22.

and its diagonals. By making different selections of the diagonals, a variety of designs results. The second figure shows a parabola traced by its tangents. In order to draw this, one takes two intersecting lines and steps off equal line segments on each line (the segments on the two lines do not need to be equal, but they usually are). Connect any one of these points on one line with any one of the points on the other, then move from point to point on one line going toward the point of intersection, on the other moving away. It is not necessary to have the point of intersection the end point of any line segment, but it usually is, on each line. The third diagram shows a curve of pursuit. Start with a curve, l; a point, P, not on l; and a point Q on l. (In our diagram, l is a straight line but it need not be). Then let $QQ_1 = Q_1Q_2 = Q_2Q_3$, and P_1 lie on PQ, P_2 lie on P_1Q_1, P_3 lie on P_2Q_2, so that $PP_1 = P_1P_2 = P_2P_3$. Then these lines outline the curve which a dog, starting at P would take if he always aimed at a rabbit starting at Q who runs along curve l at a constant rate of speed. Try l as

SENSORY LEARNING APPLIED TO MATHEMATICS 137

a circle and P as its center; investigate the effect of changing the ratio between the speeds of the two. Have pupils formulate pursuit problems and try them out in the schoolyard. Introduce colored thread to stitch artistic designs. Investigate the backs of the cardboard where the stitching is done to discuss the economic ways of carrying over the thread. A good deal of excellent mathematical thinking will result.

Improving Attention by
Controlling the Stimulus

Example 26: The stimulus may be controlled by controlling its intensity or its contrast. The presentation of geometric forms as intersections of curves is possible; when these curves are presented in striking black and white diagrams such as those in Fig. 23 it is

Fig. 23.

impossible not to be attracted to the diagram and to search for the geometric forms described (1: 64–81).

Example 27: In Figure 24 we see a demonstration that a trajectory is a parabola. The wooden stick has strings attached at equal distances; the lengths of the strings are proportional to the squares one, four, nine, etc. At whatever angle the stick is held the beads at the ends of the strings assume the form of a parabola.

138   THE LEARNING OF MATHEMATICS

Fig. 24.

This is an interesting statement to prove (30: 43, 45). The stream of water from the nozzle is kept in a constant position and the bead parabola fitted to it. Then it will be seen that the angle at which the stick is held is the same as that of the nozzle. Of course, this takes some time to plan and set up. But I defy any teacher to plan this demonstration with his pupils and to put it on in front of the class without discovering that change and movement in the stimulus will command the attention of the pupils. Mathematics will pick up in interest from that day on.

Improving Attention by Controlling the Observer

Example 28: Everyone has seen a pendulum swing back and forth in a clock, or seen a stone swinging on the end of a string. The interest of most individuals will be caught by the question, "What is the function connecting the length of a pendulum and the time it takes to swing back and forth?" Long and short pendulums, from two inches to forty feet, could be set up, and measurements made. There will be plenty of interest and attention, and finally the figures will yield the results:

$$T = k\sqrt{\bar{L}}$$

SENSORY LEARNING APPLIED TO MATHEMATICS

Later this may be made more sophisticated by showing that

$$T = 2\pi \sqrt{\frac{L}{g}}$$

but this is not necessary.

Example 29: Everyone knows about mirrors and will be caught by the experiment of trying to find a formula connecting the angle in degrees, A, at which two mirrors are set, and the number, N, of images of a light, L, arranged as in Figure 25. It is wise to make

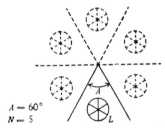

$A = 60°$
$N = 5$

Fig. 25. Light reflected in folding mirror.

the original assumption that A is an exact factor of 360. It can be then found inductively that

$$N = \frac{360}{A} - 1$$

It is important in this demonstration to vary the angle between the mirrors and stop whenever another image is fixed. Start with $A = 180$.

Example 30: Soap bubbles are familiar objects and will keep pupils' attention until the mathematics is extracted (30: 97–99). In Figure 26a we have a wire ring with a handle; inside is a loop of thread attached to the ring in three places. If this is placed in a soap solution and the soap film inside the loop broken by the corner of a piece of paper then the loop will take the form of a circle as shown in Figure 26b. This is a dramatic and convincing demonstration that the circle is the greatest area with a given perimeter. The reasoning goes like this: The surface tension in the soap film tends to make that area as small as possible; since the wire frame is fixed, the area of the shape of the loop is made as large as possible. Of course this is not a proof, but it is convincing.

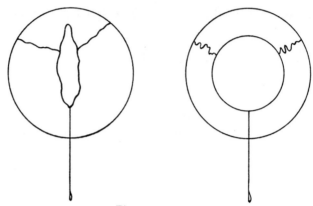

Fig. 26 a and b.

It catches that interest and attention, which is a prerequisite to learning the mathematics.

Improving Perception Through Better Organization of the Stimulus

Example 31: The perception of the number of objects in a group is very important. One of the most important factors making this perception easy is the type of organization in the stimulus. One set of examples will be seen in Figure 27. Flash one

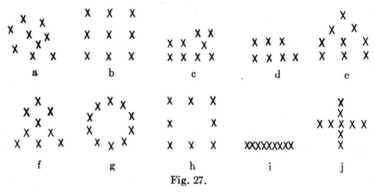

Fig. 27.

of these before an individual and increase the time of presentation until the correct number in the group is perceived. Then ask him to tell how he determined the number. Here are some of the typical responses received when this was tried with college students:

(a) 1, 2, 3, 4, 5, 6, 7, 8, 9
(b) 3, 6, 9 *or* 3 times 3 is 9
(c) 4 and 5 is 9
(d) 2, 4, 6, 7 *or* 3 and 4 is 7
(e) 2, 4, 6, 9 *or* 6 and 3 is 9
(f) 4, 7, 9 *or* 4 and 5 is 9
(g) 1, 2, 3, 4, 5, 6, 7, 8, 9, 10
(h) 9 minus 1 is 8.
(i) 2, 4, 6, 8, 9
(j) 4 2's plus 1 is 9 *or* 4 and 5 is 9

Here is a field for more fascinating research in perception of number groups.

Example 32: A familiar example of organizing the stimulus to achieve better perception is the use of colored chalk in drawing plane geometry figures. By drawing corresponding sides of congruent or similar triangles in the same color, by shading in areas to give them a unity, by keeping the coloring of the diagram in pace with the accomplished part of a proof we can add much to the perceptions organized and abstracted from perceptions.

Improving Spacial Perceptions in Mathematics

Example 33: The development of the concept of ordinal numbers is essentially one of spacial relations applied to mathematics. This is evident in the common use of visual number scales to represent the ordinal scale and the extension of this number scale to negative numbers.

Example 34: It is difficult to answer the question, "What is the locus of a point on the circumference of a circle which rolls without slipping on a straight line?" Few get this correct the first time and many will not believe the answer unless they can actually see the curve formed as in Figure 28 by rolling a wooden circle

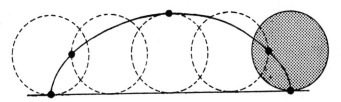

Fig. 28.

along the chalk tray and letting a piece of chalk which projects through a hole in the circumference trace on the blackboard.

Improving Temporal Perception in Mathematics

Example 35: In discussing problems of temporal perception a great deal of mathematics can be learned, and the improvement of temporal perception will improve the learning of mathematics. Reaction time experiments (27: 207–18) are a good illustration. Have a row of pupils line up each with a pencil in one hand. Each one watches the pencil of the person at his left and raises his pencil when he sees that raised. The time for a signal to pass from the left to the right of the line is determined and, by dividing by the number in the line, we find the average simple reaction time. This can be compared with the simple reaction time for auditory and tactual sensations. Then the discrimination reaction time and the choice reaction time might be investigated. The ideas sound complicated but they are not, since we need not enter into psychological explanations of "why" or "how." Arithmetical calculations will have more meaning when attached to such perceptions.

Improving Perception of Movement in Mathematics

Example 36: It is somewhat difficult to see the path of a rapidly moving projective and to analyse its movements visually. A mathematical discussion and justification of the parabola will sharpen the perception of all trajectories thereafter (1: 75–76). We carry away from a situation what we bring to it. In Figure 29

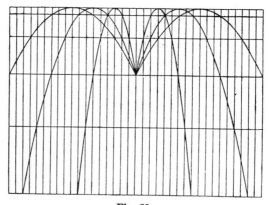

Fig. 29.

we see a graphical way of explaining the parabola; The horizontal distances are equal because we shall assume that any initial push in that direction remains constant; the vertical distances are in proportion to the squares (1, 4, 8, 16, etc.). This latter sequence is explained thus:

t	0	1	2	3	4	5 $\cdots$	t $\cdots$
a	$2k$	$2k$	$2k$	$2k$	$2k$	$2k$ $\cdots$	$2k$ $\cdots$
v	0	$2k$	$4k$	$6k$	$8k$	$10k$ $\cdots$	$2tk$ $\cdots$
s	0	k	$4k$	$9k$	$16k$	$25k$ $\cdots$	t^2k $\cdots$

If the acceleration is constant, $2k$, then the velocity at any time, t, is proportional to the time, $2kt$, and the distance, s, is proportional to the average velocity:

$$s = \left(\frac{0 + 2tk}{2}\right)t = t^2k$$

By changing the amount of horizontal push, different parabolas may be obtained. Graphical experiments with initial vertical pushes, and horizontal winds which give acceleration can be carried out. This is altogether a much more satisfactory way to perceive motion of falling particles than complicated experiments with photography or electrical sparks.

Improving Motivation in Mathematics

Example 37: We have said previously that motivation is aided by clarifying the goal and many of our examples should, therefore, have demonstrated the use of sensory learning in mathematics to improve motivation. However, here are two more. The concept of probability is one of the easiest to motivate through references to bridge hands, shooting craps, lotteries, roulette, one-armed bandits, and other games of chance. Another concrete (i.e. sensory) example for motivation is Buffon's needle problem (15: 246–47). A uniform needle of length L is dropped upon a plane surface ruled with lines H units apart, where H is greater than L. If an experiment is performed and the ratio of the number of times the needle crosses a line to the number of times it is dropped is computed, it will be found that this value approaches the number $\frac{2L}{\pi H}$. It is not necessary to prove this fact to appreciate the fact that the theory of probability summarizes the results of experiments which may be very long or difficult to carry out. Hence one is motivated to study the earlier, more accessible topics of probability.

Example 38: A familiar problem, but one which is always capable of motivation is the 64 = 65 problem shown in Figure 30.

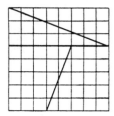 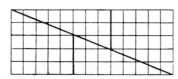

Fig. 30.

If a square is ruled into 64 smaller squares and cut as shown, and then the four pieces rearranged into a rectangle which is 5 by 13 units, we seem to have shown that there are now 65 small squares. The interest and the subsequent computation of the angles at the corners of the rectangle by simple, right-triangle trigonometry, serves as an excellent motivation device for that subject. Because physical materials have been drawn, cut and rearranged, one feels that this is a convincing proof. The triumph of mathematics over the senses in final conviction is no less important than the initial feeling, due to the sensory presentation, that this is a problem worthy of attention.

Improving Memorizing in Mathematics

Example 39: Memorizing still is, and always will be, important in learning mathematics. Recent attacks have rightly criticized meaningless memorizing. After meaning is established, drill is necessary. Many recent devices in the form of games have been produced for arithmetic teachers. Figure 31 shows a number of these. The important point is that meaningless repetition of number facts, even though using a large variety of the senses, and those very intensely, will be of little use compared with the playing of meaningful games during, and after, the significance of each number fact is established.

SENSORY LEARNING APPLIED TO MATHEMATICS 145

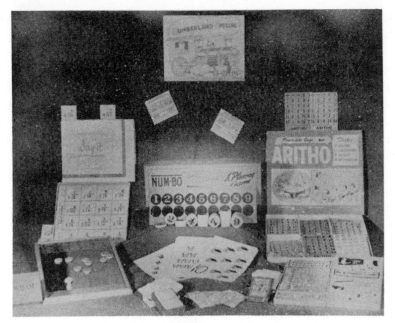

Fig. 31.

Improving the Understanding of Concepts in Mathematics

Example 40: Many times have we stressed the fact that concepts, clear in the mind of the teacher, cannot be transferred in that form into the minds of the pupils. A slow and ingenious presentation of examples and applications of the concept must be given in terms of concrete sensations so that the mind of each individual pupil can build his own concept. A good example is the concept of large numbers. Can you imagine 2 objects? Also 10 objects? How about 100? What happens to the meaning of the concept 34,567,398,682,905,562 when we try to interpret it in our imagination as a collection of objects? Recently a radio commentator, in mentioning the size of the current budget of the United States government tried to help his radio audience by telling them how many dollar bills this was for each individual in the United States, how long it would take to count all the dollars if one counted one per second, how high a pile it would make if piled up, how long a strip, if the dollars were laid end to end, how big an area it would cover if the dollars were laid out, and how heavy a pile of bills this would make. Here is an excellent series of appeals to sensory perceptions, in terms of familiar sensations, to give meaning to the concept of the very large number of dollars. The success of the establishment of this concept depends upon the amount of familiarity the audience has with the counting, size, and weight of dollar bills, and also with the power of the imagination of the audience to extrapolate these sensations to bigger amounts.

Example 41: The concept of complex roots to cubic equations is complicated enough so that all sensory aids to its understanding are worth considering. In Figure 32 we have a method

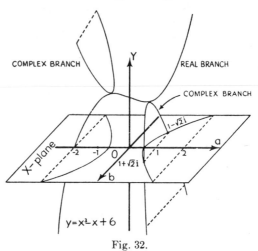

Fig. 32.

of constructing such a visual sensation to strengthen the concept for the equation $y = x^3 - x - 6$. The figure is composed of a horizontal complex x-plane and a real y-axis. The graph in the x-plane is a hyperbola and consists of all values of x which are complete complex and which give real values to y. These real values of y lie in the right hyperbolic cylinder whose elements pass through the hyperbola in the x-plane. These y values form the two complex branches shown. By appending the graph of y for real values of x, we have the total graph and the complex roots shown in the figure. Even though this explanation is complicated and needs more detail (10:288–91), it would be much more so without the visual model to accompany the explanation. A tactual model would be even better.

Improving Problem-Solving in Mathematics

Example 42: One of the most effective ways of presenting a problem situation is through sensory learning. Too often the problems for solution are verbalized and generalized in an abstract form before the pupils get them. The important steps of realizing that there is a problem in the situation, of formulating that problem, and of sloughing off the excess information are seldom done by the students. Here is an example of how sensory learning can present such a problem. Stand an ordinary platform balance from the physics or chemistry laboratory on the desk (30: 28). Rock it back and forth and show that the surfaces always remain horizontal. Ask the class why? After a proper amount of time, they should be able to draw a diagram similar to Figure 33. This shows

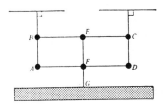

Fig. 33.

the mathematical statement of the situation. We can show the need to define the word "horizontal," which is not usually part of geometry, as perpendicular to "vertical"; and the latter as a line through a given point and the center of gravity of the earth. If the base of the balance is horizontal and AB and CD are equal, and AD and BC are continuous pieces also equal; and if EFG is continuous so that there are pivots at A, B, C, D, E, and F, we can prove that the top two platforms are also horizontal. The polishing of the statement of this problem from the physical situation and its proof may take as long as the proof of ten problems which have been presented in more abstract form, but the former experience is, at times, much more valuable and should certainly not be excluded.

Improving Emotional Learning in Mathematics

Example 43: The auditory sensations are often neglected, so here is an example using them. Various piano chords, first with two notes, then three, and possibly more, should be played for the class. In each case they should write down whether the combination sounds pleasant or unpleasant. Absolute agreement between members of the class is neither expected nor important. In fact, the degree of pleasantness or unpleasantness might be discussed on the basis of the votes. Combinations which are clearly considered to be pleasant or unpleasant by a majority should be analysed further by introducing the frequencies of the notes. The ratios of these combinations should turn out to be simpler for pleasant combinations than for unpleasant ones (5: 339). Such relationships between the arts and the sciences may open the doors to other interests for many pupils in the class.

Example 44: If we wish to approach the question of beauty in design by means of regular repetition of geometric forms, one way to do this is to talk about the kaleidoscope. This leads to the problem of repeated reflections in a number of mirrors (7: 162–64). Since a plane mirror is the perpendicular bisector of each line segment connecting a point and its image, we are immediately at the mathematical statement of the problem. In Figure 34 we see

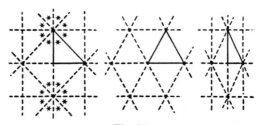

Fig. 34.

the results of reflecting a point in four mirrors forming a square or in three mirrors forming a triangle. The regularity of these figures is responsible for both the geometric beauty and the geometric simplicity. The diagrams on paper are not enough. Pupils should be encouraged to try combinations of mirrors and a lighted candle and then to justify mathematically the images seen.

Improving Imagery in Mathematics

Example 45: The improvement of memory images has been considered in connection with memory; we shall here consider the improvement of imagination images. One characteristic of such imagination images is their transitional nature. There are many mathematical concepts, the majority concerned with limits and infinity, which have this characteristic also. The imagination images connected with definitions in the calculus of tangent to a curve, area under a curve, volume of a solid of revolution, length of a curve, and many other elementary concepts require a certain amount of imagination to see the limiting process as it enters into the definition. Diagrams in books and on the blackboard help; motion pictures would help even more. There are, however, physical phenomena which exhibit this transitional nature also. One interesting example is the formation of Chladni figures on a vibrating plate (30: 187). By clamping a metal plate tightly at the center and causing it to vibrate at the edge we get such figures as that seen in Figure 35. It is necessary to sprinkle sand over the plate before it starts to vibrate and to hold the plate at some point to make a node. The sand will collect where there are nodal lines and be thrown off where the plate is vibrating. By continuing to bow and to shift the position of the finger which is making a nodal point we get one figure changing into another. It is easy to say that the plate is vibrating in sections which can be determined mathematically and which are usually symmetric, it is often necessary to help the imagination by seeing the result in sand.

Example 46: Our main argument here has been that sensory learning is a prerequisite to more abstract concepts. There are some people who also enjoy moving in the other direction in their imaginations—they will start with mathematical abstractions and express them in concrete, sensory form. One of the most delightful attempts at this has been in the books by Lieber and Lieber, for example *The Education of T. C. Mits* (20). At first the drawings seem merely entertaining and piquant, but when viewed in connection with the text, which itself is disarmingly entertaining, we see a visual and imaginative interpretation of mathematical ideas which clarifies as well as amuses. It is as impossible to catch the appeal and explain it to justify the right word in

150 THE LEARNING OF MATHEMATICS

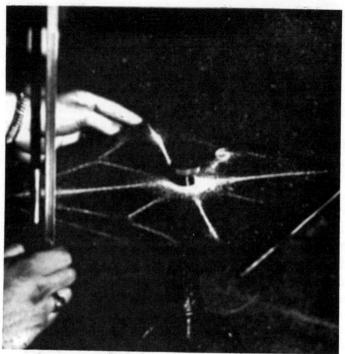

Fig. 35.

Shakespeare or the appropriate phrase in Beethoven; one element of imagination is its spontaneity and elusiveness. There is far too little of it in mathematics; that may be one reason our subject is sometimes considered cold. The example, Figure 36, is completely inadequate without the accompanying text and fellow drawings It is merely an invitation to see the original.

Example 47: Some mathematical concepts defy the imagination. How would you draw a curve which becomes infinite in length but is closed, continuous, and surrounds a finite area? Figure 37 illustrates a curve which does this. By continuing the process begun in these diagrams, we get a curve of the desired type as the limit (15: 343–55; 29: 83–85). Is not the connection between sensory learning and imagination clear? How could one imagine the pathological answers to the questions posed above without the visual presentation of the method by which these curves are con structed? Of course, a verbal explanation and description of this

SENSORY LEARNING APPLIED TO MATHEMATICS 151

Fig. 36.

process is possible, but one need only try to formulate it to see the cumbersome and confusing substitute which it is for a sensory presentation.

Now we have finished our attempt to give specific, concrete examples of how sensory learning can be employed in the teaching of mathematics to improve aspects of that teaching which we all desire. We have applied this to the use of visual sensations; auditory sensations; olfactory and gustatory sensations; cutaneous, static, kinesthetic, and organic sensations; and to the use of muscular skills; also to the improvement of attention by controlling the stimulus and the observer; to improving perception by better organization of the stimulus; to the improvement of spacial perception, temporal perception and the perception of movement in mathematics; and to the improvement of motivation, memoriz-

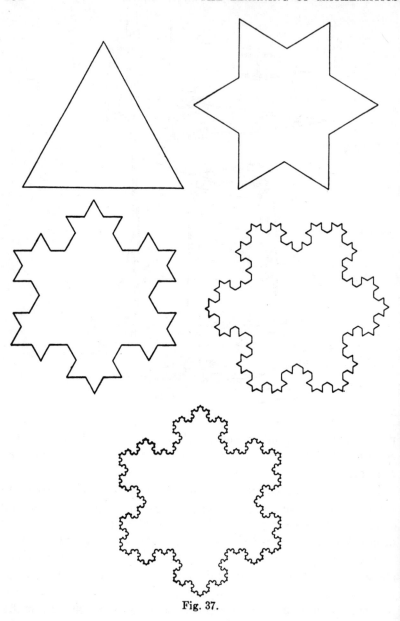

Fig. 37.

ing, understanding of concepts, problem-solving, emotional learning, and the formation of imagination images. These are the contributions of sensory learning to the teaching of mathematics.

Previously we had shown how mathematics could improve sensory learning and had used as our examples there the development of motor skills, improvement of attention, perceptual discrimination, and perceptual illusions.

In order to place this whole discussion in the right framework we began with a short summary of the concepts of sensation, attention and perception, and the relationship of sensory learning to higher types of learning such as motivation, memorizing, concept formation and use, problem-solving, and emotional learnings.

May we end on this note? Sensory learning is not everything, but it is very, very important.

Bibliography

1. BARAVALLE, H. V. "Geometric Drawing." *Multi-Sensory Aids in the Teaching of Mathematics.* Eighteenth Yearbook. National Council of Teachers of Mathematics. New York: Bureau of Publications, Teachers College, Columbia University, 1945. p. 64–81.
2. BEAUMONT, HENRY, and MACCOMBER, F. G. *Psychological Factors in Education.* New York: McGraw-Hill Book Co., 1949.
3. BORING, E. G.; LANGFELD, H. S.; and WELD, H. P. *Psychology.* New York: John Wiley, 1935.
4. BROWN, WARNER, and GILHOUSAN, H. C. *College Psychology.* New York: Prentice-Hall, 1950.
5. BURNS, E. E.; VERWIEBE, F. L.; and HAZEL, H. C. *Physics, A Basic Science.* New York: D. Van Nostrand Co., 1943.
6. COLE, L. E., and BRUCE, W. F. *Educational Psychology.* New York: World Book Co., 1950.
7. COURANT, RICHARD, and ROBBINS, HERBERT. *What Is Mathematics?* London: Oxford University Press, 1941.
8. CRUZE, W. W. *General Psychology for College Students.* New York: Prentice-Hall, 1951.
9. DOUGLAS, O. B., and HOLLAND, B. F. *Fundamentals of Educational Psychology.* New York: Macmillan, 1938.
10. FEHR, HOWARD F. *Secondary Mathematics, A Functional Approach for Teachers.* Boston: D. C. Heath and Co., 1951.
11. GALTON, FRANCIS. *Inquiries into Human Faculty and Its Development.* New York: Macmillan, 1883.
12. GARRETT, H. E. *Psychology.* New York: American Book Co., 1950.
13. GIFFORD, W. J., and SHORTS, C. P. *Problems in Educational Psychology.* New York: Doubleday Doran, 1931.
14. JUDD, C. H. *Educational Psychology.* Boston: Houghton-Mifflin Co., 1939.
15. KASNER, EDWARD, and NEWMAN, JAMES. *Mathematics and the Imagination.* New York: Simon and Schuster, 1940.

16. KELLEY, W. A. *Educational Psychology.* Milwaukee: Bruce Publishing Co., 1947.
17. KINGSLEY, H. L. *The Nature and Conditions of Learning.* New York: Prentice-Hall, 1946.
18. KRAITCHIK, M. *La Mathématique des Jeux.* Bruxelles: Stevens Frères, 1930.
19. LAWTHER, J. D. "Development of Motor Skills and Knowledge." *Educational Psychology.* (Edited by Charles E. Skinner.) New York: Prentice-Hall, 1946. Chapter 7.
20. LIEBER, L. R. *The Education of T. C. Mits.* New York: Norton and Co., 1944.
21. MCCAMMAN, C. V. "Curve-Stitching in Geometry." Eighteenth Yearbook. National Council of Teachers of Mathematics. New York: Bureau of Publications, Teachers College, Columbia University, 1945. p. 82–85.
22. MURSELL, J. L. *The Psychology of Secondary School Teaching.* New York: Norton and Co., 1932.
23. NATIONAL COUNCIL OF TEACHERS OF MATHEMATICS. Eighteenth Yearbook. *Multi-Sensory Aids in the Teaching of Mathematics.* New York: Bureau of Publications, Teachers College, Columbia University, 1945. p. 379–95.
24. NIKLITSCHEK, ALEXANDER. *Im Zaubergarten der Mathematik.* Berlin: Scherl, 1939.
25. PERRY, WINONA. *A Study in the Psychology of Learning in Geometry.* T. C. Contributions to Education, No. 179. New York: Bureau of Publications, Teachers College, Columbia University, 1925.
26. RUCH, F. L. *Psychology and Life.* Chicago: Scott, Foresman, 1948.
27. SEASHORE, C. E. *Elementary Experiments in Psychology.* New York: Henry Holt and Co., 1908.
28. SHUSTER, C. N., and BEDFORD, F. L. *Field Work in Mathematics.* New York: American Book Co., 1935.
29. STEINHAUS, H. *Mathematical Snapshots.* New York: Oxford University Press, 1950.
30. SUTTON, R. M. *Demonstration Experiments in Physics.* New York: McGraw-Hill Book Co., 1938.
31. TENNYSON, J. A. "Window Transparencies." *Multi-Sensory Aids in the Teaching of Mathematics.* Eighteenth Yearbook. National Council of Teachers of Mathematics. New York: Bureau of Publications, Teachers College, Columbia University, 1945. p. 86–87.
32. TROW, W. C. *Introduction to Educational Psychology.* Boston: Houghton-Mifflin Co., 1937.
33. WARREN, H. C., and CARMICHAEL, LEONARD. *Elements of Human Psychology.* Boston: Houghton-Mifflin Co., 1930.
34. WOLFF, GEORG. "The Mathematical Collection." Eighth Yearbook.

National Council of Teachers of Mathematics. New York: Bureau of Publications, Teachers College, Columbia University, 1933. p. 216–43.
35. WORCESTER, D. A. "Memory by Visual and by Auditory Presentation." *Journal of Educational Psychology* 16: 18–27; 1925.
36. WUNDT, WILHELM. *Mythus und Religion, Völkerpsychologie.* Vol. II, Part I. Leipzig: Wilhelm Engelmann, 1905.

5. Language in Mathematics

IRVIN H. BRUNE

WHAT IS LANGUAGE?

LANGUAGE can help and language can hinder learning. Intelligent living requires that we transmit thoughts; we communicate by means of language. Through language man has shared his discoveries, widened his understandings, preserved his learnings, developed his civilizations, and educated his children. Thus language has benefited mankind. Yet, because at best it reveals meanings imperfectly, language has produced misunderstandings, bred dissensions, and even fomented wars. The power of language, like the force of fire, can effect good or ill in human affairs.

In the teaching of mathematics language has also both succeeded and failed. Whenever it has led pupils to enjoy the satisfactions of thinking through a mathematical situation, language has helped. Whenever it has engendered lack of clarity as pupils seek to solve problems, language has hindered. In this chapter we shall consider a few principles of communication, and make a few suggestions to teachers. We shall aim to help teachers and pupils scrutinize language. It will be a terse treatment, but even a modest study of language helps us to perceive the power of words. In the drama of thinking, language plays the lead.

Suppose that we consider examples:

1. An experienced teacher merited recognition. For 30 years she taught mathematics to boys in a correctional school—a place where one tries hard to salvage youthful deviates. Of course the teacher sought no honors, but some of her friends gave a dinner for her. When pressed for the secret of her success, she spoke neatly: "The boys and I have always understood that *right is right*."

2. A pupil in algebra also spoke neatly: "Irregardless of its sign a number squared gives a positive result."

3. Another pupil erred often by making hasty generalizations "All generalizations," he said, "are false."

LANGUAGE IN MATHEMATICS 157

4. A girl said, "The more I correct my arithmetic, the worse it becomes."

Each of the four foregoing illustrations reveals a weakness in language; we note that such language hinders communication.

The teacher's words pleased her friends, but the words meant little. We who seek to improve our own teaching get no specific help from the truism "right is right."

The pupil meant well, but double negatives in English follow a grammatical rule similar to the mathematical rule he had in mind about negatives. The word "irregardless" upset his message; he simply didn't say what he thought he was saying.

The other pupil spoke profoundly, but his statement couldn't hold. "All generalizations are false" includes itself, and therefore denies itself.

The girl corrected her arithmetic; hence it couldn't become worse. She, too, unwittingly upset her statement by her own words.

Here are examples of another sort:

5. "Arrange six toothpicks to form four equilateral triangles." Mary proposed this puzzle.

6. Jim, a third-grader, remarked that "Cokes cost 7¢ apiece, so you can get three cokes for a quarter and have 4¢ left."

7. Tim in Grade II said that "there are 3 tens in 30, and one ten has 2 fives, and so 30 has 6 fives.

8. The eighth grade defined and redefined *ratio* until they agreed that "a ratio compares by division two numbers expressing the same kind of units."

9. Sally, a ninth-grader, said, "A circle is a ring around a point. If you measure to the point from places on the ring you always get the same distance."

The last five examples show strengths of language; we see how language helps communication.

Mary with a minimum of effort entertained her classmates with a simple, clear challenge. Her words were concise, nice, precise.

Jim showed that he understood a quantitative situation and could convey his thoughts about it.

Tim handled a child's syllogism based on a thorough understanding of certain numbers.

Discussion by a group tests a statement. The eighth grade rethought and altered their words until they produced a highly refined statement.

Sally's description of a circle qualified as a good definition: it contained simple words; it put *circle* into a class; it distinguished *circle* from other rings; it was brief.

Teachers of mathematics know that mathematics says more in fewer symbols than any other language. The examples 5, 6, 7, 8 and 9 we have just examined attest to the clarity, conciseness, and precision of mathematical statements. Spitzer lists three statements that show how the language of mathematics enhances a report:

> a. During recent times some of the original soil of our cultivated slopes has washed away. (b) During the last five years two inches of our original soil has washed away. (c) During the last five years about one-fourth of the eight inches of the original soil has washed away (29).

Mathematical language facilitates thinking by complementing ordinary language (as in Spitzer's example above), and it also suggests solutions to problems. Let us look at an example. Measurements have shown that gold loses about one-nineteenth of its weight if it is weighed in water rather than in air. Similarly, silver weighed in water loses about one-tenth of its weight. If, then, a quantity of an alloy of gold and silver weighs 12 ounces in air and 11.16 ounces in water, how much of the alloy, weighed in air, is gold, and how much is silver? The power of mathematical symbolism appears when we let n represent the number of ounces of gold, weighed in air, in the alloy. Then $12 - n$ represents the number of ounces (weight in air) of silver in the alloy. And the equation $\frac{18}{19}n + \frac{9}{10}(12 - n) = 11.16$ both states concisely the conditions of the problem and suggests clearly how to solve the problem. As Sol Worth explained to Matt, his younger brother, "To solve an equation, you undo it." The results, $n = 7.6$ and $12 - n = 4.4$, represent ounces of gold and silver in the alloy.

Mathematics as a way of thinking concerns itself with language. Mathematics begins with terms understood and accepted without

LANGUAGE IN MATHEMATICS 159

definition by all concerned. "Number," "counting," "same," "equals," "plane," "distance," "point," "path," are among those often thus used. Using undefined terms as key words, students of mathematics then agree on certain definitions. A *locus*, for example, is the "path" of a "point" moving according to certain accepted conditions. If the "path" is restricted to a "plane" and if the "path" must everywhere be the same "distance" from a chosen "point" in the "plane," then the *locus* may be given a name, such as *circle*. Further assumptions, often called postulates, are then agreed on. These statements can be linked together to form proofs. If, for instance, we agree that (a) equal circles have equal diameters and (b) areas of circles vary as the squares of their diameters, then we must agree that (c) equal circles have equal areas.

This kind of reasoning lies at the heart of mathematics. The conclusions, of course, come from the assumptions. Obviously, faulty language can addle the process. When meaning eludes us, correct conclusions can evade us too.

A few examples indicate how language stymies conclusions:

10. "Like signs give plus." Pupils glibly say these words, and reach wrong conclusions in algebra.

11. Mr. Parfit joined a cult and testified that "the new order changed the course of my life 360°." Somewhere, somehow language led Mr. Parfit to a ridiculous conclusion.

12. A junior high-school pupil averred that he could solve a problem about ages, "by mathematics but not by algebra." Language apparently threw him for a loss here.

13. Usually no fewer than 90 per cent of a high-school class will choose an income of $4000 annually with an annual increase of $200 as preferable to an income of $2000 semiannually with a semiannual raise of $50. Language makes the worse proposition seem to be the better.

14. A lad argued thus: If the seventh problem is worked correctly, the answer is 12. I got 12 for the answer when I worked the seventh problem. Therefore I worked the seventh problem correctly. Here plausibility of language replaced validity of reasoning.

Mathematics deals with clear, consistent, concise, and cogent language. But language does not always help to the degree in which it is mathematical. The statement "Riemann's straight line is endless, but finite" is consistent. Yet few high-school pupils understand it. Its meaning doesn't transfer because the pupils lack experience with Riemannian geometry. Bertrand Russell's "The number of a class is the class of all those classes that are similar to it" (28) met out of context also puzzles pupils. The pupils again lack background; they aren't ready linguistically. For language succeeds only when meanings transfer. The sender must say what he means in words that suggest to the receiver what the sender had in mind.

Accordingly, students of language count any device for transmitting meaning as language. Signs, gestures, spoken words, written words—all serve as symbols to suggest and recall meanings. When Jack surreptitiously spells out words to be seen only by fellow members of the Beaver Patrol he deals in language just as surely as a senator does when he puzzles people with polysyllabic profundity. So also Jack uses language when he senses a problem, states it, translates it into equations, and resolves it by solving the equations. Further, the Zulu beats out messages on tom-toms, and the florist distributes cards listing the language of flowers. All people use language of some sort to exchange ideas.

The primary purpose of language, then, is to convey meanings. Any device which does so is language. And we wish to make our language a benefit, not a detriment; it should help, not hinder.

WORDS AND MEANINGS

Words are links in the chain of communication. We remember reading in Longfellow's poem that Paul Revere arranged with his friend: "One, if by land, and two if by sea." We know that fire, smoke, drums, pistol shots, dots and dashes, red and green lights, and the like, convey messages. We realize that a glance can speak; a shrug may mean much. We note, too, that non-verbal signals usually derive their meanings from verbal agreements. Paul Revere knew what *two* lights meant because he and his friend had made a *verbal* agreement. Red and green lights originally were "stop" and "go" lights; the words were clearly visible against the colors. Words underlie all thinking and exchange of thought.

Words represent agreements among people. They avail much—if they do not baffle.

Spoken words are symbols. They represent people, objects, acts, and many things, much the same as a snore in a sound film signifies sleep. Just as the snore, moreover, is not sleep itself but merely a symbol for sleep, so also words are not reality itself but merely symbols for reality. Arbitrary sounds call life-facts to mind.

Written words represent spoken words. Hence the words one sees on a printed page are symbols for symbols. The process of using symbols for symbols may be extended. The words "John's height" may represent sounds which designate a physical actuality. But one may use h to denote John's height. To designate John's, Bill's, Susan's, and Mary's heights one may write: h_1, h_2, h_3, h_4, respectively. These symbols of symbols of symbols enhance brevity, clarity, and efficiency. One may substitute yet another symbol for the symbols $h_1 + h_2 + h_3 + h_4$ so that further conciseness results:

$$\sum_{i=1}^{4} h_i.$$

Thus a single symbol may stand for the results of an operation performed on many quantities. The symbols of mathematics are greatly condensed, generally unambiguous, and easily manipulatable. Economy in time and effort results.

Teachers need to recall that mathematical words often represent mental constructs rather than tangibles. One does not experience pure number or geometric magnitudes through one's senses. "Five" represents a property common to $\genfrac{}{}{0pt}{}{xxx}{xx}$ and /////. Two "planes" meet in a "line"; two "lines" meet in a "point." Yet points, lines, and planes are ideas. "Point," "root," "circle," in mathematical usage, represent abstractions. We note that in one sense lines, though one-dimensional, consist of zero-dimensional points. Again, one-dimensional lines are intersections of two-dimensional planes. No serious student of words takes words to be bits of reality, and mathematical words often do not refer at all to reality. They symbolize fictions—mental constructs.

These observations do not gainsay the values of teaching and learning aids. Rather they emphasize those values. The more abstract an idea is, in fact, the more children need to work with concrete devices. Intellectual activity necessarily deals with abstractions. Perceiving an idea in many real-life situations helps the pupil make abstractions himself. When Billy Primary sees four apples, four children; handles four crayons, four sticks; draws four rings, four tallies; counts four papers, four pencils; and hears four notes, four blasts; he can note for himself that *four* is not people, objects, or marks; it is a group just so big.

Let us now look at other mathematical words.

Suppose, for example, that John Curious, a junior high-school pupil, encounters the word "root." This word represents a sound which in turn represents a thing. In accordance with his experience to date, the thing "root" suggests to John is that part of a tree which grows underground and fixes, supports, and nourishes the tree. For him the phrase "root of an equation" may seem farfetched. If, to date, his life in school has encouraged his natural curiosity and not, as is often the case, stifled it through memorizing and verbalizing, John may seek to get at the root of this word "root." Did he not hear in Sunday school that "the love of money is the root of all evil"? Why did the teacher of Latin recently refer to the Latin root of the English word "percentage"? Clearly, a tree successfully transplanted, roots itself firmly in the earth. Is this related at all to the mild upheaval of earth occurring when a hog is said to root? Don't cheer-leaders exhort all pupils to root for the team? What has all this to do with the root of an equation? "Come to think about it," muses John, "we found square roots of numbers last year. Is the root of an equation a square root?"

The teacher who really helps pupils with language in mathematics will go beyond the minimal requirements of the situation in which the words "root of an equation" are to be taught. Merely to tell John that a root of an equation is a known quantity which satisfies the equation when substituted for the unknown quantity in the equation will not suffice. Such a statement is concise, correct, and clear—to an adult versed in mathematics. But to John, whose experiences with mathematical language are slowly spiraling and spelling meanings for him, an adult's pat phrases and neat formulations may be practically meaningless. However, if his instructor requires textbook talk and/or teacher talk, John can memorize it. Better still, he can receive credit for his memorizing unmeaningful language; he can, in fact, if he memorizes well, even win school honors and rewards from his parents.

How, then, should the teacher help pupils to understand the word "*root*"? In the first place pupils should have much experience with many equations and should have discovered for themselves that in some cases a single value satisfies the equation and in other cases two or more solutions can be found. In discussing and summarizing their work, pupils will probably need a word which they can use to distinguish the answer(s) to an equation from other kinds of answers they obtain in mathematics. Then, and no earlier, should the name be given. The *idea* that one or more values satisfies an equation is important. After it has been discovered, the naming of such values is an easy matter. *Pupils who get the idea and then name it tend less to become confused than pupils who attempt to learn terms representing ideas which are as strange as the terms themselves.*

Another way in which the teacher can help pupils to understand the word "*root*" is to discuss some of the word's many meanings. He can encourage his pupils to pool their knowledge of the word. He can direct their attention to possible relationships among the various meanings. Is there some similarity of meaning among the phrases, the supporting part of a tree, the basic part of a word, the root of a matter, the root of all kinds of evil, the root of a dam, the root of a tooth, the fourth root of 256, and the root of $x/3 + 27 = 54$? Are these in turn related at all to the ideas of becoming fixed, or implanting, or causing roots to grow? Where too does the idea of eradicating come in? To help pupils use language intelligently teachers must take the time to contrast mathematical meanings with non-mathematical meanings of words. Fortunately mathematical words usually are very simple in form.

In ordinary speech and writing, *usage* indicates what words mean. In mathematics careful *defining* sharpens word meanings. For example, "run" may mean to scamper, to flow, to smuggle; or "run" may be a stream, a score in baseball, or a disaster in hosiery. But mathematically "run" is the horizontal component of the length of a rafter, or it is a distance covered in a period of time.

Instances of words which represent technical as well as ordinary meanings abound. "Root" has already received some comment in this chapter. Accuracy, precision, error, base, carry, borrow, check, difference, product, foot, common, concrete, round, prime, order, principal, rate, reciprocal, solid, representative, sieve, significant, similar (the list could go on and on)—these are ordinary words which take on precise meanings when used in mathematical settings.

Multiplicity of meaning we naturally expect. No two people have exactly similar backgrounds of experience. We learn words in diverse settings, and hence no two people can associate entirely identical values with a given word. To understand one another at all, we have to fix meanings of words arbitrarily. Usage, more or less agreeable to all, settles meanings—at least for a time. And the technical words of mathematics, even though they may resemble ordinary words in form, are more rigidly defined than ordinary words.

Further light on the understood-by-agreement nature of words comes by studying shifts in the meanings of even technical words. We suggest only a few: "remainder" in subtraction as compared with "remainder" in division, square "root" versus "root" of an equation, "hypothesis" as a part of a geometric theorem as opposed to a hypothesis to be tested by experiment, "statistics" as facts on the one hand and as a procedure on the other. Even "sum" refers to a subtraction that Alice did for Humpty Dumpty (21).

A billion, moreover, means a thousand million to an American. To a Briton, though, a billion means a million million. A perfect language probably would embody the principle of one word for one idea. To date no such language is in use, and mathematics is no exception. Of course mathematics is relatively precise in its use of words and symbols, but even the simple $19 \times 8 \div 4 \times 2$ can be ambiguous. In such matters the teacher cannot aim merely to be understood. He must express himself so clearly that he cannot be misunderstood.

Another way to sensitize pupils to the arbitrary nature of words seldom receives the attention it deserves. Most pupils respond readily to occasional references to foreign words. One can request them to select, for example, the best word from the following list: carré, tetragonas, Viereck, square, cuadrado, quadra. Except for pupils who know one or more of the foreign words in the list, the selection of course will be the word *square*. The point then to be brought out is that to a German the best word would be *Viereck*, to a Frenchman, *carré*, and so on for the other word in the list. Indeed the word is not the object or the idea, but only its accepted symbol.

Teachers mindful of the arbitrary nature of words readily recognize such pitfalls of language as the beliefs that:

1. Speakers and writers unerringly choose exactly the words they need to express their ideas.
2. Listeners and readers receive meanings exactly as they were transmitted.
3. Any given word has a unique meaning.
4. Children who state the correct answer necessarily understand the concept(s) in question.

A person's ability to understand words doubtless indicates probable success in the complex life of our times. The efforts of makers of intelligence tests to measure such ability bear witness to this relationship. Yet, when by dint of much reading, hearing, and saying, a pupil merely mimics and mouths others' words—when he resorts to verbalisms—he probably progresses but little toward maturity in thinking. A pupil who recites glibly how to factor the difference of two squares and then proceeds to evaluate $87^2 - 13^2$ by squaring and subtracting reveals empty verbalizing rather than understanding and applying a mathematical idea.

Words are tools, forged in the experience of the race, and sharpened by agreements among people. Yet individuals use them; and individuals may communicate clearly, or they may let themselves be misunderstood. Let us recall the conversation between Humpty Dumpty and Alice:

"... that shows that there are three hundred and sixty-four days when you might get un-birthday presents."

"Certainly," said Alice.

"And only *one* for birthday presents, you know. There's glory for you!"

"I don't know what you mean by 'glory,'" Alice said.

Humpty Dumpty smiled contemptuously. "Of course you don't—till I tell you. I meant 'there's a nice knock-down argument for you!'"

"But 'glory' doesn't mean 'a nice knock-down argument,'" Alice objected.

"When I use a word," Humpty Dumpty said, in rather a scornful tone, "it means just what I choose it to mean—neither more nor less."

"The question is," said Alice, "whether you *can* make words mean so many different things."

"The question is," said Humpty Dumpty, "which is to be master—that's all" (21).

SEMANTIC

Old Man Memorizit is dead, as dead as a fossil.

Teaching is more than telling and explaining, and learning is more than imitating and memorizing. During the last 60 years teachers of mathematics have gradually sensed that, above all else, their pupils should learn the *meaning* of mathematical terms, principles, operations, and patterns of thought.

Mary wasn't bothering to understand geometry. Ten years of schooling led her to believe that memorizing must be the best way to earn passing grades. Of course the definitions, assumptions, theorems, and proofs of geometry were novel, but her study habits weren't. So she completed the test item, "A circle is _____," by writing "a clothed curve." Mary's memory had tricked her.

Jack memorized a rule that one may "cancel like factors in the numerator and the denominator of a fraction." So in a quiz in trigonometry he wrote $\frac{\cos x}{\cot x} = \frac{\not{c}\not{o}\not{s}\,\not{x}}{\not{c}\not{o}t\,\not{x}} = \frac{s}{t}$. Jack had carried mechanization too far.

Cross-multiply was a kind of magic to Frederick. The sight of fractional coefficients was a sort of quick-trigger situation for him. Hence $\frac{x}{3} = \frac{x}{4} + 5$ set him to writing $4x = 3x + 5$ and $x = 5$. Did he check his result? No. "A nice whole number like 5 must be right." Frederick had manipulated without understanding.

Sally was furious. She had reduced $\frac{a^2 - b^2}{a + b}$ by writing $\frac{\not{a}^{\not{2}} - \not{b}^{\not{2}}}{\not{a} + \not{b}}$ and then by writing $a - b$, and Miss Blanck herself had later read $a - b$ as the correct answer. Sally also had fallen for meaningless manipulations.

Perry skipped all stuff about logic. So to him this reasoning was valid:

If an official is honest, he will reduce taxes. Mr. Mayor reduced taxes during his first term. Therefore, Mr. Mayor is honest.

Also Perry felt that since every equilateral triangle is isosceles, then every isosceles triangle is equilateral. Perry should have investigated converses.

It shocked these pupils to learn that they were notoriously out-of-date. In colonial days pupils memorized their teachers' words, learned the rules by heart and wrote exercises neatly into their copybooks. Since the turn of the twentieth century, however, the trend has been toward thoroughness, understanding, self-reliance, and thinking. Pupils nowadays observe, count, measure, estimate, solve, check, and reason. They solve sensible situations—live problems, not busywork.

Indeed, busywork is passé—nearly dead. In 1945 the Commission on Postwar Plans of the National Council of Teachers of Mathematics summarized what teachers believe. The Commission wrote:

We must give more emphasis and much more careful attention to the development of meanings . . . it is a mistake to accept glib verbalism as evidence of sound learning.

Meanings do not just happen. Nor can they be imparted directly from teacher to pupils, as by having them memorize the language patterns in which meanings are couched. Instead, meanings grow out of experience, as that experience is analyzed and progressively reorganized in the thinking of the learner. In a word, each child creates his own meanings; accordingly teacher activities are perforce restricted to those of guidance. It is the function of the teacher to provide an abundance of relevant experiences and to assist the child to isolate the critical elements and to build them into the desired understandings (9).

Nowadays teachers aim to develop thinkers. Teachers know that, although words bear close watching, yet words are thinkers' tools. So teachers contrive situations in which pupils discover mathematics. Pupils see for themselves that three and five are eight, that there are four sevens and two left in thirty, that a cubic inch is a kind of measuring stick, that fractions can be figured out with diagrams, that Euler's diagrams test many an argument, that measuring an angle precisely is usually easier than measuring a line precisely, that division by zero is impossible, and so on through years of growth in mathematics. And teachers search day by day for adequate words. They need words to describe problem situations, words to question pupils' unreasoned statements, words to encourage further pupil research. Teachers hunt words to relight the flame of curiosity which incited pupils before pat phrases and answers-in-alabaster smothered it. Teachers plan daily to challenge, to ask why, to doubt, to interest, to evaluate, and to exploit numerous ways to further pupils' growth. Good language challenges: it does not bore; it does not frustrate. And teachers help pupils to handle words:

Men use words to solve most of their perplexities, if not all of them. But it is not easy to use words properly in solving problems.... In helping students to think reflectively, therefore, the teacher should help them to understand the use of words.... (26).

Just about the time when leaders in mathematics education began to question memorized mathematics, Michel Bréal began to question lax language. In 1900 he wrote *Essai de Semantique*. From it we got the word "semantics," the study of words and meanings.

Bréal noted that certain words (he used "compressibility" and "immortality" as examples) contain all that the idea contains (5). "Height," for instance, expresses completely one and only one idea. So also with "five"; the word and the idea coincide.

Bréal noted also that other words symbolize objects, and these words cannot suggest all the ideas which the objects suggest. In mathematics, for example, "triangle" calls up a plane figure, a spherical figure, or a pseudospherical figure. Each of these could be scalene, isosceles, or equilateral. Moreover, a plane triangle could be rectangular, a spherical triangle could be rectangular, birectangular, or trirectangular; a pseudospherical triangle could not be birectangular. The word "triangle" varies in meaning according to what we know about the figure; although actually it focuses attention on the property of three angles. Similarly, "number" suggests quantity, although we can invent kinds of numbers as we will.

Bréal saw too that words sometimes perpetuate an incorrect notion. We know now that electrons move in a vacuum tube from a relatively negative cathode toward a relatively positive plate. Originally, though, electricity was thought to flow from an abundance (positive) toward a lack of charge (negative). Other examples crop up in mathematics. "Borrow" in subtraction does not suggest what we really do. "Penny" is an English, not an American coin. "Remainder" in subtraction hardly suggests that we need 10¢ if we have 25¢ and wish to spend 35 cents.

Bréal observed further that once a term gains acceptance, its relevancy matters little. A mathematical example, "imaginary number," illustrates this point. We hold nowadays that all numbers sprang from man's imagination; man invented numbers to serve him. So-called imaginary numbers are indeed genuine, and their applications are very real. But "imaginary numbers" persist in our language.

A note of regret appeared in Bréal's *Essai*. Apparently he considered words to be bits of reality, segments of truth. Seemingly he deplored discrepancies between the face values of words and the objects the words represent. We know today that words, like money, symbolize values acceptable to their users, rather than marvel, as Bréal did, that words distort impressions. (An irrational number is by no means a crazy number.) We set it down as a principle that:

Words, as everyone now knows, "mean" nothing by themselves, although the belief that they did ... was once equally universal. It is only when a thinker makes use of them that they stand for anything, or, in one sense, have "meaning." They are instruments (23).

Granted that a speaker and a hearer are both trying to communicate and not befuddle, they symbolize, transmit, receive, and interpret meanings. An exchange of meanings does result, but the speaker's thought has small chance of carrying perfectly. A writer and a reader also understand each other imperfectly because most words suggest several meanings.

Miss Bland, teacher of mathematics, knows that writing lacks inflections, emphases, gestures—oral aids to understanding. She knows, too, that speaking often lacks the precision needed for understanding. So she talks and "chalks." Yet Miss Bland knows that telling and explaining fall short. Teaching is more than telling.

Bill, a pupil in mathematics, knows that he must understand an idea before he can symbolize it. Words and/or mathematical shorthand can mar or make meanings. Sometimes Bill would rather quit at the non-verbal stage. He knows, for instance, that a^2 means $a \cdot a$ and that a^3 means $a \cdot a \cdot a$. The expression a^n, however, he understands but would rather not verbalize. "I know it," he says, "but I can't say it."

To help people use words effectively semanticists point to principles somewhat like those in some paragraphs that follow. Like other students, semanticists disagree on some matters. There are schools of semantics; there are issues. But the principles we mention find rather general agreement.

1. *A Way of Life.* Semantics, a young discipline, connotes at present an attitude rather than a science. Illustrative cases abound, but as yet scientific data and scientific laws aren't numerous in this field. The goal, of course, looms. Semantics bids to become the science explaining how *language affects other behavior, especially thinking.*

To date, "consciousness" summarizes the subject in one word. People who realize that language has power, people who apply semantic principles, people who think before they speak, people who recognize that language both tells *and* excites—such people are conscious semantically.

2. *Context.* We have noted, in the section on words and meanings, that use and use alone determines what a specific word means. How it is used and in what milieu of other words it appears set its meaning. A car may be "fixed," for example, the better to win a race; but a horse may be "fixed" to lose a race; a boat may be "fixed" to move but little—to ride at anchor; and a point may be "fixed" in space to move not at all. Surely the meaning of

"fixed" is not uniquely fixed. In each of these cases, then, the idea conveyed by the word "fixed" comes from the context built up by it and other words conjointly. The meaning of a word is known, let us say, by the company the word keeps—by its context.

The pupil in arithmetic early encounters technical words which vary according to context. From situations such as: "John had 15 cents, and he earned 10 cents more. How much money did he then have?" the pupil may associate "more" with addition. When the same pupil meets the circumstance: "John has 15 cents, but the movie he desires to see costs 25 cents. How much more does he need?" then the pupil finds the word "more" in another context; the question, he notes, requires subtraction. "Forevermore" is yet another "more."

Similarly, "remainder" in one context is the answer to a subtraction, whereas "remainder" in division is a somewhat different concept. Examples of mathematical words which shift in meanings according to context include: hypothesis, proof, zero, postulate, pencil, decimal, degree, order, average, induction, statistics, majority, pair, accuracy, range, curve, power, tangent, and square.

With respect to context the point for teachers to keep in mind is that pupils should learn that nuances (precise shades of meaning) and multiple definitions matter much—that the meaning of a word depends on how it is used.

3. *Science.* Teachers of mathematics are seldom in a position to guide pupils through the experiments of natural science, or to adduce evidence to substantiate the details of scientific subject matter, or to teach formally the organized materials of science. A sound basis for teaching mathematical application, however, includes awareness of the scientist's point of view:

a. Scientists report events. They seek to discover what happens in the world, and they abstract principles which explain events. The behavior of scientists at work is as important as the body of knowledge which accrues from their observations. Above all else scientists *perceive* a number of instances of principles before they *enunciate* those principles. It is a kind of behavioral understanding preceding their verbalization of results. Teachers who encourage pupils to discover mathematical principles by counting, measuring, and experimenting are using a powerful procedure. Pupils, like scientists, thereby report events after they have behavioral understanding of them. They know what they did to get specific results. They generalize on the basis of first-hand experience.

b. Matter is a process—a whirling of electrons. Hence, objects are really events, and change is everywhere the rule rather than the exception. People, objects, and relationships actually change from instant to instant. Naturally problems arise. With the flow of events new ideas, new relations, new understandings, and new solutions ensue. The teacher of mathematics helps pupils to see changes, to cope with new problems, and to abstract principles which apply to events, which, though they resemble one another, never exactly duplicate one another. Thinking, of course, helps people adjust to change. New occasions teach new duties. People gain new insights, and people solve new problems.

c. A concept is a set of operations. A person best understands length by measuring lengths; a pupil who has counted the square inches within the confines of a closed figure understands the concept of area; and one who has carried a 30-pound pack appreciates the concept of weight. For pupils learning mathematics those questions unanswerable by actual operations tend to be meaningless. Again, as in items a and b in this section, understanding appears as behavior. What the pupil does reflects what the pupil understands.

d. Besides discovery and explanation, prediction and verification interest the scientist. In situations requiring problem-solving, pupils can be encouraged to study the problem, hazard a guess as to its answer—predict the outcome—and then gather and analyze data to verify or to refute the prediction. Such a procedure ties the so-called scientific method in with mathematical situations.

4. *Abstracting.* Teachers reflect the attitude of scientists when they keep certain principles in mind as they work with their pupils. One who recognizes the process character of reality, who notes that change is the essence of existence, and who appreciates the complexity of even simple events will tend toward scientists' modesty and scientists' tolerance rather than toward conceit and dogmatism.

Every event has an infinity of characteristics. To describe an event people necessarily select, usually unwittingly, what they feel is important in the event. They simply cannot include all its properties. They report the salient features. They necessarily abstract relatively few items and base their statements on them. It is not possible to recount exactly "the truth, the whole truth, and nothing but the truth."

Teachers of mathematics need to be conscious of abstracting. It, as the keystone, supports the arch of semantic principles. Teachers need to realize, and help their pupils to realize, that whereas reality changes and gets complicated, statements about reality oversimplify and persist. Herein lies the value of inductive procedures. Teachers encourage their pupils to find a real problem that the pupils solve by counting, measuring, recording data, and generalizing results. Class discussions enable each pupil, each a budding scientist perhaps, to put into words the items he abstracted from the problem. Pupils exchange observations they made in the situation. They eventually agree, under the teacher's guidance, upon properties common to a variety of problems. They abstract properties, relations, and probably conclusions. They experience, abstract, verbalize. Observation, generalization, and communication derive from abstractions.

Pure mathematics, of course, establishes general relations among abstractions. Pupils measure the sizes of angles in several triangles not merely to know about particular angles in specified triangles. Their teacher guides them to discover, i.e., to understand behaviorally properties and relations among angles in all plane Euclidean triangles. The word "plane" is an abstraction, being a mental rather than a physical construct, and it limits the field of investigation. The word "Euclidean" limits the discussion still more because it denotes and connotes a particular set of postulates abstracted from a multitude of possible geometric assumptions.

Teachers who help their pupils to understand abstracting are helping them not only to discover mathematical principles but also to communicate those principles to their classmates. Pupils work with projects specifically designed to emphasize mathematical concepts, they discover the concepts in the situations, they understand them from their experiences with them, and they state in words the ideas they have gleaned. The child who abstracts "three" as a property common to certain collections of objects and who has learned "five" by handling things, can discover for himself what three and five together are, and he will find words for telling others about his discovery. Similarly, the pupil who constructs right triangles, measures sides and angles, and computes ratios between lengths of sides will discover properties of similar triangles and properties of trigonometric ratios. Having, moreover, behavioral understanding, he can proceed to verbalize and submit his report to his classmates for discussion, rewording, and acceptance. Abstracting from actual experiences thus provides a foundation for deductive procedures, the goal of mathematical endeavor.

5. *Referents.* Most people, if confronted with the question: Is a word the thing it represents? would probably reply hastily that it certainly is not. Yet many of these same people behave as if words were things, as if names were realities, as if symbols were referents. To be called a subversive is not to be a subversive, although the appellation excites people as much as a spotlight disturbs a burglar plying his trade.

Teachers of mathematics deal with a subject in which the terminology is relatively emotionally sterile. Yet those teachers realize that the people they deal with matter more than the subject they teach, and that, for their pupils, mathematical words may have emotional overtones. For some youngsters "fraction" frightens, for others "division" distresses, for some "variable" vexes, and for many "algebra" is Arabic.

We need not multiply illustrations; teachers can easily cite many more instances of words which wreak wonderment and worry. Teachers also note from the behavior of their pupils that mathematical words can become blocks to learning. Who has not sensed a tenseness among pupils when a word such as "ratio," or "coefficient," or "inversion" is heard?

To analyze such verbal hurdles in the path of the learner helps more than to list them. Why do some pupils shudder when they encounter a word like "variation"? One main reason is that sometimes pupils learn words as abstractions devoid of concrete referents. To them the word is the thing. Yet, on the other hand, the pupil who measures, records, and compares heights and weights of his classmates, can note and understand variation. The pupil who counts the square inches of surface within rectangles of different sizes and shapes can understand how area varies jointly with length and width. The pupil, however, who merely hears about variation, direct variation, inverse variation, joint variation, and variation as the square or as the cube of an independent variable, can fail to catch the idea of "variation." If the pace necessary to cover a course of study has to be at all accelerated as far as the *learner* is concerned, the idea of "variation" may be quite unclear. If the pupil tries to memorize verbal distinctions among direct, inverse, and joint variation without getting the feel experimentally —without behavioral understanding—he may eventually get lost and become frustrated. Eventually the word "variation" may incite tenseness and block learning.

Another reason why technical words may be bugbears to some pupils is the very precision of those words. A person, as long as he avoids unsocial acts, may have vague notions about words such as "truth," "brotherhood," "honesty," "virtue," "wisdom," and so on, without particular harm to himself or others. One usually does, in fact, have one's own ideas about such words. For a mathematical word, however, fuzzy notions are inadequate. It seldom suffices to know that "mean," "median," and "average" connote central tendency. Judgments as to which is the better, the mean or the median, depend on one's knowing exactly how each is defined. As another example, the pupil considering space may refer to the *middle* of the land, and be unspecific. When he refers to the *center* of a circle, however, he is specific. To deal loosely with these words, to use them interchangeably, say, is to encounter semantic difficulties.

The fact that users of mathematical words may shift meanings in different contexts has already been mentioned. Such shifts, of course, occur less frequently for mathematical words than for ordinary language, but they can trouble pupils who confuse words with referents. If he takes the word to be actually the thing, the pupil who has worked long in a workshop to make a transit for field work finds only nonsense in the sentence, "The ferry makes 10 transits a day." If he thinks symbols are objects, he may wince at the appearance of the word "radical" because political, chemical, mathematical, and literal connotations of the word are a confusion to him.

In studying mathematics pupils have a unique opportunity to note that there is no necessary relation between a word and its referent. Words are more or less useful conventions; "a rose by any other name would smell as sweet"; an unretouched skunk by any other name would smell. Meanings are arbitrary; users make words mean what they want them to mean. Teachers in mathematics may well emphasize this point. Perhaps words cannot be made emotionally sterile. They can, nevertheless, be regarded in their true light—as servants, not as masters.

6. *Reactions.* Human nervous systems cooperate through symbols—mostly speech and writing. What one learns one conveys to another, who reworks it and uses it. When people act in accordance with information, when they base decisions on evidence, when they prove propositions, when they evaluate others' thinking, they depend on symbols. They think matters through, they weigh meanings, they react to symbols.

Such reactions require at least a slight pause for reflection. When Cookie took her homework to Dad for help, they learned to divide 11 apples among 4 people by working it out with apples. They considered the apples representing the symbols describing the problem, and got 2 whole apples for each person. Then they reacted further to the symbols and cut the remaining apples into fourths.

Symbol actions necessarily are delayed; we react, not to the symbol itself, but to its referent. Naturally we perceive the symbol before we recall its meaning. We need time to find and test that meaning. Cookie and her dad found meaning in the problem about apples as they handled some apples. They did not fret because Cookie couldn't give an automatic answer. They didn't time their work. Rather they *took time* to understand, they got at the meaning, they solved the problem, and they had no regrets.

But there are other reactions. The day following the experience with apples, Cookie's dad phoned home and said he was bringing a surprise. When he entered the house, two children and six dogs immediately charged him. What was it? What's the surprise?

Disaster ensued. The cream puffs Dad carried yielded to the pressure, and spattered at large.

Automatic, little pondered, or habitual acts are signal reactions. Such reactions require no time for thinking; they are practically undelayed. Korzybski emphasized the relation between reactions and thinking: "The symbolic levels are uniquely human and differentiate most sharply *human* reactions from signal reactions of lower, less complex forms of life" (19).

When Cookie, her brother, and their dogs leapt upon cream-puff-laden Dad, their reaction was somewhat animalistic. For, although lower forms of life react to signals, the converse (signal reactions remain subhuman) does not follow. Many human acts proceed from almost instant recognition of meanings. When Junior learns to drive the family car, he practices many of the operations until they become almost automatic. In mathematics too, Junior solves enough equations (after he has discovered the principles) to routinize the work. He responds to a stop light and he responds to $.20(5 - n) + .40(n) = .25(5)$ quickly. In either case he saves time and energy for matters that require reflection.

Helping pupils to blend signal reactions and symbol reactions wisely requires teachers' best efforts. Convinced that knowing instantly what eight sevens are pays, pupils may seek to mechanize problem-solving. If they desire a number to express success in making baskets in basketball, they may subvert thinking and clutch at cues. In the phrasing "Twenty-seven baskets is what per cent of 56 tries?" they may divide 27 by 56 "because 56 goes with of." If the same problem comes to them in the words "Lotto set a league record of 27 baskets in 56 tries last night, what per cent of success was that?" they may divide 56 by 27 *"because 27 goes with of."*

In geometry pupils may strive to acquire a set of signal reactions. And the going is arduous. To memorize one proof after another reaps small satisfaction. Confronted by a proposition they have studied from a book or from their teacher's words, pupils may try to repeat the proof from memory. If they succeed they may get good marks; if they slip they fail miserably. If, on the contrary, pupils approach a theorem as a problem, they will react to symbols. What do the assumptions mean? What words have we defined? What follows from the facts at hand? The pupils try to discover their own proofs; they look for meanings, link them together, and test results. They react to symbols.

How we react matters. Signal reactions facilitate operations, once we know why they work. Symbolic reactions facilitate thinking, problem-solving, proving, and evaluating the propositions people offer us.

7. *Maps.* Diagrams, charts, graphs, scale drawings, and maps depict relationships. They symbolize some abstractions we have made from real situations. Hence they represent territories somewhat as words represent ideas, as languages represent cultures, and as numbers represent quantities. Korzybski likened the relations between symbol and referent to the relations between map and territory in this manner:

A. A map may have a structure similar or dissimilar to the structure of the territory.

B. Two similar structures have similar logical characteristics. Thus, if in a correct map, Dresden is given as between Paris and Warsaw, a similar relation is found in the actual territory.

C. A map *is not* the territory.

D. An ideal map would contain the map of the map, the map of the map of the map, endlessly.... We may call it self-reflexiveness.

Languages share with the map the above four characteristics.

A. Languages have structure...

B. If we use languages of a structure non-similar to the world and our nervous system, our verbal predictions are not verified empirically, we cannot be rational or adjusted...

C. Words *are not* the things they represent.

D. Language also has self-reflexive characteristics. We use language to speak about language... (18).

Numerous interpretations and illustrations of the foregoing principles appear in Korzybski (18), Rapoport (27), Johnson (16), Hayakawa (13), and Keyes (17).

We shall mention only three examples, however, to illustrate that, since words, maps, charts, drawings, and diagrams represent reality incompletely and imperfectly, people sometimes err in trusting them implicitly. No map shows all details; map makers rely on measurements, all of which only approximate actual lengths; territories change, and maps become obsolete. Besides overlooking imperfections in maps, users often confuse their own inferences with descriptions the map maker wrote.

Jerry, our first example, entered college believing that a mathematician is one who can glance at Union Station and tell you how many bricks it contains. Jerry's map for exploring mathematics came partly from his own experiences and partly from others' words. Years before, when he entered Grade VII, people used the name "mathematics" instead of "arithmetic." Throughout junior high-school mathematics and in two courses in senior high school the pupils only computed. Jerry's map showed him that mathematics means lightning calculating. In fact, his father, who figures lumber quickly at the Builders' Supply Company, considers himself to be "quite a mathematician."

Miss Steofan, teacher of mathematics, uses her special kind of map. To her the beautiful conclusions in mathematics, the practical usefulness of mathematics, and the correctness of it all brought her to the belief that in mathematics people meet Truth itself. Her pupils, accordingly, confront a kind of heaven-hell dichotomy. All who learn and abide by mathematical laws, as Miss Steofan interprets the laws, reap right answers and high marks. All who doubt and deviate from the divine pattern simply fail. There is disciplinary value, Miss Steofan believes, in pupils' doing as they are told. Doubters could later become subversives.

Miss Cherie, a teacher of a fourth grade, encourages the children to bring clippings to school daily. The children seek news stories containing numbers. If stories with pictures and numbers appear, so much the better. Each child then tells the others about his clipping, writes and reads the numbers he found, and reports new words to the class. If no one in the class can help with either new words or large numbers, the pupils start research. They name the places in the numbers, the teacher helping only after all reach a digit they cannot name. They look words up in children's dictionaries. Later they read one another's clippings, make up problems about them, solve and check the problems, and then post the clippings for comparison with future reports. Miss Cherie's map of beliefs leads her to link reading, arithmetic, current events, and telling time together. Her chart also leads the class to daily periods for studying language and arithmetic systematically because these subjects are systems.

People construct word maps from direct experience and from verbal reports they hear and read. We have merely touched the topic through three illustrations. The map we ourselves are developing in this chapter indicates that Jerry and Miss Steofan might well examine their mathematical maps. All arithmetic is mathematics, but all mathematics is not arithmetic. Mathematical conclusions follow inevitably from mathematical assumptions, but man made those assumptions—they are not truth itself.

From time to time man has doubted, experimented, and invented. Mathematics grew that way. Miss Cherie's teaching map lets her deviate from lessons in a book. Her pupils look things up, discuss them, and include them in problems they compose, solve, and check. Miss Cherie is willing to doubt, experiment, and invent. She helps pupils to evaluate and build up their own word maps in ordinary and mathematical language.

8. *Understanding.* A child might easily learn by heart the following stanza from the *Pirates of Penzance.* He might commit it and then sing it. Or he might sing it over and over and thereby learn it. Either way, though, he might not understand what he says or what he sings.

> Though counting in the usual way,
> Years twenty-one I've been alive,
> Yet reckoning by my natal day,
> I am a little boy of five!

When pupils learn without seeing the point in the material, they merely memorize meaningless words. They use symbols adroitly without understanding them. They deal in verbalisms. They manipulate symbols, but think little. Korzybski had this principle in mind when he wrote: "Only the technical interplay of symbols, to find out some new possible combination, can be considered as low-grade thinking" (18: 69).

Verbalistic learning plagues pupils in mathematics as in other subjects. Pupils sometimes use numerals without understanding numbers. Frequently pupils working verbal problems seek cue words that disclose, so they believe, which operation to use. This they do without trying to understand the problem. The words "cancel," "transpose," "collect," "invert," "cross-multiply," and "simplify" also often become rather empty verbalizings to pupils. Teachers of mathematics can readily add examples to the ones that appear here.

The moral, of course, is that teachers should emphasize that facility with symbols often differs from understanding symbols. And understanding ties symbols to life. The Harvard Committee suggested:

> Abstractions in themselves are meaningless unless connected with experience.... The teacher can do a great deal ... ; he can relate theoretical content to the students' life ..., and he can deliberately simulate in the classroom situations from life ... he can be persistent in directing the attention of the student from the symbols to the things they symbolize (12).

Betz took a similar position: "The evidence is overwhelming that when mathematics is taught as a *cumulative system of ideas, with due regard for understanding and mastery, and for life-centered applications*, it ceases to be a meaningless game" (2).

Some ways that teachers can use to help pupils to substitute understandings for verbalisms follow:

Larry learned to say "nine" whenever he saw 5 + 4. Following the summer vacation Larry's new teacher took stock of his knowledge of combinations. At that time he said *"eight,"* then *"eleven,"* for 5 + 4. Besides, he didn't know how to find the correct sum.

Suggestion: Encourage Larry to count five classmates, then four classmates, and then five and four classmates. Repeat with chairs, with books, with pictures, with marks. Help Larry to experience *five* and *four* enough to understand five-ness, four-ness, and nine-ness. Have him explain orally to his classmates what the symbols 5, 4, +, and 9 mean. Have him explain orally how to find other simple sums. Have him tell often what to do when he forgets a combination. Have him prove to others that the results he offers are correct.

Claire chose $1/12$ in the following question: Which is the largest fraction: $1/2$, $1/8$, $1/4$, $1/3$, $1/12$, $1/6$?

Suggestion: Have Claire cut 1-inch strips of paper each 12 inches long. Have her cut one strip into halves, another into fourths, another into eights, a strip into thirds, one into sixths, and one into twelfths. Encourage her to tell how a fraction gets its name. Help her to discover that a large denominator suggests many parts, and that each part has to be small. Claire can also see from the pieces that the number of parts is the "namer," or denominator.

Joe wanted to find out what part of his state's population (about 2,500,000) lived in his home city (about 15,000). He wrote: $\frac{15,000}{2,500,000} = \frac{15,\cancel{000}}{2,5\cancel{00},\cancel{000}} = 5 | \frac{15}{10} = \frac{3}{5}$. When questioned, Joe could not decide whether his answer fitted the problem; he maintained that "zeros above" and "zeros below" could be "canceled"; he stated glibly that "the outside 5 in the last step '*guzinta*' 15 three times *and* 25 five times."

Suggestion: With one-inch strips 12 inches long cut into halves, quarters, eights, thirds, sixths, twelfths, fifths, and tenths, encourage Joe to show that: $\frac{2}{4} = \frac{1}{2}$; $\frac{4}{8} = \frac{1}{2}$; $\frac{5}{10} = \frac{1}{2}$; $\frac{3}{6} = \frac{1}{2}$; $\frac{4}{12} = \frac{1}{3}$; $\frac{3}{12} = \frac{1}{4}$; etc., until he can state the general principle that dividing *both* terms of a fraction by the same number (not zero) changes the form but not the value of the fraction. Ask Joe to reverse the process—show that $\frac{1}{6} = \frac{2}{12}$; $\frac{1}{3} = \frac{2}{6}$; $\frac{1}{2} = \frac{5}{10}$—and discover the principle of multiplying both terms of the fraction by the same number. Have him try his rules on $\frac{10}{20}$, $\frac{100}{300}$, $\frac{150}{250}$, $\frac{150}{2500}$, $\frac{15000}{2,500,000}$.

Allen attempted a puzzle which read as follows: "The value of a certain fraction is $\frac{3}{4}$. If one is subtracted from the numerator and if one is added to the denominator, the value of the resulting fraction is $\frac{2}{3}$. Find the original fraction. Allen concluded that "the puzzle is inconsistent; 1 from 3 leaves 2, and 1 plus 4 makes 5; so the resulting fraction should be $\frac{2}{5}$."

Suggestion: Allen was the most capable boy in his class. Yet one phrase in the puzzle meant nothing to him in that context. Word-by-word reading, and the question, "Would a shorter wording state the same thing?" eventually helped Allen. Equations did not elude him, but simple language did.

Luco cut 3 wires for bracing a television aerial. The aerial was to be mounted on a flat portion of a roof, and the wires were each to be fastened on the roof 5 feet from the base of the aerial and on the aerial 10 feet from its base. Luco recalled the Theorem of Pythagoras as "the sum of the legs squared equals the square of the hypotenuse," so he computed the hypotenuse, added 1 foot each for fastening the wire, and cut the wires. They were too long, of course, and Luco wasted some wire.

Suggestion: Luco didn't vizualize the Theorem of Pythagoras, or he would have seen the flaw in his ready rule. Models and diagrams, plus a proof of his own, would have put meaning into the symbols Luco tried to memorize without understanding them. He would have thought from symbols to squares and back to meaningful symbols.

Facility in using symbols, we conclude, does not guarantee thoroughness in understanding those symbols. Machines can solve differential equations. Only human beings, though, understand and set up differential equations to solve life problems.

Teachers can direct pupils' attention from symbols to referents and back again. Abstractions misunderstood do not help us; instead they hinder.

Teachers also can help pupils to recognize how people use language to persuade or how people sometimes seek to make the worse appear the better reason. The ancient Gorgias held that "Nothing is; or if anything is, it cannot be known; or if anything is and can be known, it cannot be communicated." Pupils today need not embrace Gorgias's creed. They should, however, be aware of semantic principles.

VERBALIZATIONS

A person's vocabulary suggests his intelligence and reflects his intellectual successes. Makers of scales for measuring mental age usually depend heavily on verbal items. The reason is that people learn words; we know that words do not erupt like teeth in children's mouths.

Bruce is just learning to talk. He can say clearly "Dad," "Mama," "Paw-paw" (for Grandpa), and "bike." Parents, relatives, and friends repeat words, and Bruce tries to imitate them in his own inimitable way. He is gradually learning to associate definite sounds with specific people and things. He points, hears, then points again, and says the sound he just heard. He is a great repeater.

Although no one to date has told Bruce so, he not only builds his vocabulary, but he also practices what logicians call "extensional defining." To date, of course, experiences with his parents have made up his social life almost entirely. When he points and says "Dad," the word means, to him, many experiences he has had with *Dad*. When he shies from strangers (i.e., people that suggest no experience—meanings), clings to his mother, and cries "Ma-ma," he again associates experiences with a word.

Indeed, when Bruce points to his dad, he offers not at all to share verbally with others the abstractions he has made about "*Dad*." He points, and lets others make their own abstractions. The same goes for "Mama," when Bruce shuns strangers and reaches toward his mother. Other people simply have to make their own abstractions.

As Bruce grows older his fund of meanings for the few words he now knows will increase. His stock of words will also increase, and the more experiences he has, the more words he will learn and the more meanings he will tie to each word he makes his own. The sounds of the words, of course, he will learn by imitation. But the meanings he will forge for himself in that lively fire of his many experiences.

Still later, Bruce will learn to see words in manuscript, in print, and in cursive. Eventually he will learn to sound out words for himself from the printed or written form. He will learn to pronounce words by associating standard diacritical marks in a dictionary with specific sounds. For meanings of words, though, he will forever depend heavily on experiences.

Bruce, in time, will learn, at home, at Sunday school, at school, at work, and at play, to tell about words by using other words. As with extensional definitions, which he uses unwittingly of course, Bruce will also eventually learn to use what logicians call "intensional definitions"; he will define words entirely by using other words. He will become an intellectual.

We hope for Bruce, though, that his parents, his leaders, and his teachers will contrive literally millions of excursions, adventures, and experiences for him. We hope that he will retain—even increase—that lively curiosity he now possesses. We hope that no efficiency-bent teacher(s) will delude him with words—entice him to seek worn words, pat phrases, catchy slogans, and smooth sentences that he can substitute for first-hand learning by doing, experiencing, and thinking. We hope that he will crave to discover things and tell and write about them in his own words. We hope that no teacher will assign words for him to tell back or write back. We do not want him to become a "mental parrot." And Bruce represents any one of millions of children who now are curious and eager to learn.

Learning words without first-hand experiences to bring out meanings hinders thinking. In verbalisms Korzybski saw beyond educational damage to possible neurological damage also:

> ...first order empirical facts are more important than definitions or verbiage. It should be noticed that the average child is born extensional, and then his evaluations are distorted as the result of intensional training by parents, teachers etc., who are unaware of the heavy neurological consequences (18: xv).

Verbalisms have abounded in mathematics—"Invert the divisor and multiply"; "Crossmultiply"; "Cancel"; "Transpose"; "Reduce"; "Bring down"; "Drop the per cent sign and move the decimal point two places to the left"; "Annex the per cent sign and move the decimal point two places to the right"; "Factor completely"; "Double the width, double the length, and add"; "Divide the number following *is* by the number following *of*"; "Add the number of decimal places in the multiplicand to the number of decimal places in the multiplier"; "Subtract the number of decimal places in the divisor from the number of decimal places in the dividend, adding zeros to the dividend if necessary"—such statements often lack meaning for the pupils who have learned to recite them. Teachers of mathematics can readily adduce more examples of verbalisms. They realize, too, that although a pupil's words may conceal his lack of understanding, yet his words may not substitute for understanding. Sooner or later the parrot quits mathematics.

To help pupils to increase their mathematical understandings teachers plan experiences for the pupils. In arithmetic, for example, a program of extended, unhurried concept-building helps pupils discover relationships among numbers before the pupils learn technical words to express those relationships. John and Mary learn to recognize groups, to count, to measure, and to solve simple problems largely through discoveries they make with groups of people, toys, sticks, pictures, drawings, and marks.

Gradually, moreover, pupils realize that the idea of combining one group with another group needs a name. The word "adding," being needed, sticks. When pupils later encounter situations that adding will resolve, the pupils have the idea and reverbalizing the idea readily follows. Experiences also uncover the ideas pertaining to other operations, and the pupils' understandings and vocabularies grow simultaneously.

Similarly in algebra pupils think through many cases of combining gains (positive integers) with other gains. Also pupils think through many cases of combining losses (negative integers) with other losses. Then they handle many cases combining gains with losses. From these experiences pupils get the idea of algebraic addition. Once they get the idea pupils understand why technical language and rules of operation fill a need. When ideas precede verbalizations, then empty verbalisms do not appear. The pupils, in fact, understand the relationships and the rules because they composed them themselves.

Similarly in geometry pupils experiment with many circles and many parallel lines before they define them. Key words in definitions should refer to thoroughly known concepts. Indeed definitions should follow, not precede, experiences. When pupils report their discoveries orally, when they express generalizations they themselves have made, when they explain solutions to problems they have solved, when they submit proofs, and when they interpret quantitative results, they grow in vocabulary. Terms needed for such reports persist, for understanding helps pupils' memories.

Pupils who really know, can communicate their knowledge. "I know it, but I can't say it" has long fascinated teachers and other students of learning. Pupils may exhibit behavioral evidence that they are grasping a concept, but they may show too that they are groping for words. Suppose we consider the statement $(a - 2)(a + 3) = 0$. A pupil may sense that a can be no bigger than 2, and he may say that he cannot tell why a is thus restricted. But teachers usually help such pupils by contriving more experiences of a similar sort. Words, crude words at first perhaps, result. But familiarity with an idea abets verbalization of it. If pupils realize that new technical words would enhance their statements, then they are psychologically ready to acquire those new technical words. And words thus learned are easily remembered.

Professor Gertrude Hendrix two years ago considerably clarified the matter of subverbal awareness as prerequisite to meaning. The excellence of her treatment, in fact, impels the present writer to cease here, and to urge all readers to seek out her article (14).

In this section on verbalizations we have contrasted verbalizations with verbalisms. We build the former on experiences, the bedrock of understanding. The latter we construct on pure memory alone, the sinking sand of meaningless words. Experiences to supplement reading, we contend, would have helped the student who a few years ago wrote this: "Things can be proved by statistics if correct data are [sic] acquired through experimental observation. But if observation is made in frustration it proves nothing."

TEACHING FOR MEANING: SUMMARY

Throughout the chapter we have suggested and implied that language profoundly affects behavior. In our case, of course, learning mathematics concerns us as desirable behavior. The pupil who understands mathematics uses mathematical expressions he encounters. Throughout their mathematical studies pupils handle symbols, recall what the symbols represent, and interpret hypotheses and deductions couched in symbols.

Pupils progress when ideas of size, degree, and relationship shine clearly to them through mathematical language. When language suggests ideas inaccurately or inadequately, however, pupils make slight headway, if indeed they make any gain at all. We have considered examples of weak language and examples of helpful language.

Adequate mathematical language is clear, concise, correct, and cogent. It complements ordinary language, and it suggests which operations to use to solve perplexing situations. It includes undefined terms, definitions, postulates, and theorems. It shows steps in a proof. It records and communicates solutions to problems.

Faulty language, however, can mislead people. A given line of reasoning may be invalid, yet it may seem plausible. Teachers of mathematics and teachers of other subjects render tremendous service to their pupils when they help them to detect the verbal booby-traps set up in modern messages sent via mass media of communication.

Words symbolize, but they are not bits of reality itself. Mathematical symbols facilitate thinking, provided pupils clearly understand what the symbols refer to. Spiral learning develops from many contacts—concrete experiences at first, and then situations progressively more and more abstract.

Printed pages record meanings through symbols; but the reader gleans these meanings to the degree that his background of experiences with those symbols permits. Meanings fixed by definition in mathematics help learners understand writers and teachers. But individual experiences inevitably affect interpretations; if symbols mean but little, pupils may memorize them to get by, or they may detest them and hate mathematics.

Originally "to teach" meant "to show." Later "to teach" came to mean "to tell." Teachers intent on efficiency practiced telling as a way to save time. Woeful deficiencies in mathematical learning, however, cropped out. To correct these weaknesses teachers retold the facts and drilled pupils on the facts. But drill devoid of understanding did not enhance learning. Pupils could operate— they could add, subtract, multiply, divide, factor, and recite proofs—but many failed to solve problems. Only the few who forged meanings for themselves understood mathematics. Their fellows disclaimed mathematical ability and despaired. Mathematical illiterates greatly outnumbered those with mathematical competence.

Nowadays teachers realize more and more that their task is to contrive experiences for their pupils to think through. Modern teachers devise ways for pupils to discover meanings for themselves. Facility with language, of course, helps pupils learn mathematical meanings, and conversely.

One key idea which teachers keep in mind resides in the word "awareness." Teachers who use language effectively realize that language makes or mars communication. They understand that context determines the meaning of multivalued words. They appreciate that, whereas words tend to be stable, the life-events that words symbolize tend to change. They measure a pupil's grasp of a concept by what he does with the concept. They are aware that reports inevitably abstract and depict only salient features from an infinity of characteristics. They reiterate the principle that words are not events—that words merely represent events. They differentiate thinking (reacting to symbols) from automatic acts (reacting to signals). They emphasize likenesses between maps and words—they note that faulty words, like faulty maps, lead us astray. They seek, above all else, understanding instead of many mere manipulations of meaningless marks.

Teachers seek to transfer this consciousness of semantic principles to situations outside the class room. Citizens in a democracy need to be wary of words. They encounter at every hand words designed for them to accept with a minimum of careful evaluation. Instead of encouraging pupils to memorize empty symbols, teachers in America should help pupils to think critically and weigh their own and others' words.

Bibliography

1. BELL, ERIC T. "The Meaning of Mathematics." *The Place of Mathematics in Modern Education*. Eleventh Yearbook, National Council of Teachers of Mathematics. New York: Bureau of Publications, Teachers College, Columbia University, 1936.
2. BETZ, WILLIAM. "The Necessary Redirection of Mathematics, Including Its Relation to National Defense." *The Mathematics Teacher* 35: 166; April 1942.
3. BLOOMFIELD, LEONARD. "Linguistic Aspects of Science." *International Encyclopedia of Unified Science*. Chicago: University of Chicago Press, 1939.

4. BODE, BOYD HENRY. *How We Learn.* Boston: D. C. Heath and Co., 1940.
5. BRÉAL, MICHEL. *Semantics.* (English Translation by Mrs. Henry Cust). New York: Henry Holt and Co., 1900. p. 171–72.
6. BRIDGMAN, P. W. *The Logic of Modern Physics.* New York: Macmillan, 1932.
7. CARNAP, RUDOLF. *Introduction to Semantics.* Cambridge: Harvard University Press, 1942.
8. CHASE, STUART. *The Tyranny of Words.* New York: Harcourt, Brace and Co., 1938.
9. Commission on Post-War Plans of the National Council of Teachers of Mathematics. *The Mathematics Teacher* 38: 200–201; May 1945.
10. DEWEY, JOHN. *How We Think.* Boston: D. C. Heath and Co., 1933.
11. FAWCETT, HAROLD P. *The Nature of Proof.* Thirteenth Yearbook, National Council of Teachers of Mathematics. New York: Bureau of Publications, Teachers College, Columbia University, 1938.
12. HARVARD COMMITTEE REPORT. *General Education in a Free Society.* Cambridge: Harvard University Press, 1945. p. 70–71.
13. HAYAKAWA, S. I. *Language in Thought and Action.* New York: Harcourt, Brace and Co., 1949.
14. HENDRIX, GERTRUDE, "Prerequisite to Meaning." *The Mathematics Teacher* 43: 334–39; November 1950.
15. HOGBEN, LANCELOT. *Mathematics for the Million.* New York: W. W. Norton and Co., 1937.
16. JOHNSON, WENDELL. *People in Quandaries.* New York: Harper and Brothers, 1946.
17. KEYES, KENNETH S., JR. *How to Develop Your Thinking Ability.* New York: McGraw-Hill Book Co., 1950.
18. KORZYBSKI, ALFRED. *Science and Sanity:* An Introduction to Non-Aristotelian Systems and General Semantics. Lakeville, Conn.: Institute of General Semantics—Second edition, 1941. p. 750–51.
19. KORZYBSKI, ALFRED. *Lectures.* Holiday Seminar, Institute of General Semantics, 1946–47.
20. LEE, IRVING J. *Language Habits in Human Affairs.* New York: Harper and Brothers, 1941.
21. CARROLL, LEWIS. "Through the Looking-Glass." New York: The Heritage Press, 38: 200–201; May 1945. p. 111–12.
22. MORRIS, CHARLES W. *Signs, Language and Behavior.* New York: Prentice-Hall, Inc., 1946.
23. OGDEN, C. K., and RICHARDS, I. A. *The Meaning of Meaning.* New York: Harcourt, Brace and Co., 1927. p. 9–10.
24. PEI, MARIO. *The Story of Language.* Philadelphia: J. B. Lippincott Co., 1949.
25. PIAGET, JEAN. *The Language and Thought of the Child.* (English

Translation by Marjorie Warden). New York: Harcourt, Brace and Co., 1926.
26. PROGRESSIVE EDUCATION ASSOCIATION. *Mathematics in General Education.* New York: D. Appleton-Century Co., 1940. p. 214.
27. RAPOPORT, ANATOL. *Science and the Goals of Man.* New York: Harper and Brothers, 1950.
28. RUSSELL, BERTRAND. *Introduction to Mathematical Philosophy.* Second edition. London, 1920. p. 18.
29. SPITZER, HERBERT F. *The Teaching of Arithmetic.* Boston: Houghton-Mifflin Co., 1948. p. 23.
30. TRIMBLE, H. C., BOLSER, F. C. and WADE, T. L., JR. *Basic Mathematics for General Education.* New York: Prentice-Hall, 1950.
31. WALPOLE, HUGH. *Semantics: The Nature of Words and Their Meanings.* New York: W. W. Norton, 1940.
32. WEINBERG, ALVIN N. "General Semantics and the Teaching of Physics." *The American Science Teacher* 7: 104–8; April 1939.

6. Drill—Practice—Recurring Experience

BEN A. SUELTZ

DEFINITION AND AGREEMENT

DURING the past quarter century the word "drill" has not only changed semantically but also, and more important, the significance of drill has changed. Twenty-five years ago, drill was the common method of learning applied to such school subjects as arithmetic, writing, and spelling. Children were required to write a word 50 times to learn to spell it and the present generation of middle-aged people spent countless minutes in winding up ovals in one direction and then unwinding them in the opposite direction in order to train the muscles to follow the sweeping curve of penmanship. This was drill, it was carried to extremes and became so sterile that during the 10-year period of approximately 1935 to 1945 drill, as a learning procedure. was frowned upon and ridiculed in many educational circles. However, during the same period it remained the dominant pattern employed by many teachers. More recently, drill, as a part of the learning process, is again respected. But it is not drill for drill's sake that we respect, rather it is its contribution to meaningful learning that aims to become functional for the individual. In order to be most fruitful, drill must be employed with artistry. This is not an easy, mechanical, or formulated artistry, but it is one that requires a high level of discernment in knowing *when*, *how much*, *where*, and *how* to apply.

In this chapter, the words "drill," "practice," and "recurring experience" are used to indicate those aspects of learning and teaching that possess elements of similarity or sameness which repeat or recur. These recurring experiences should have a commonality that is discernible by the learner. In this discussion the words "drill," "practice," and "recurring experience" will be used within the framework that some authors refer to simply as "drill." The following examples will illustrate the inclusiveness of the discussion.

1. A four-year-old boy was stacking blocks of various sizes into a column which repeatedly fell down after he had placed a few blocks upon each other. He tried holding the blocks with one hand but as soon as the hand was removed the column fell. He discovered that straightness or perpendicularity was a factor and tried to arrange them accordingly and met with better success. Then his father showed him how to place larger blocks at the bottom. The boy tried again and met with more success and showed the glow of accomplishment. Note that this experience involves not only drill or practice but also elements of discovery. The boy is an active participant. He uses a combination of mental, visual, and manual avenues of learning. He has used drill-experience in each of these avenues of learning. The intimacy of drill as a part of learning is also apparent.

2. A ten-year-old girl has a slip of paper on which she has written the number combinations "5 × 8 = 40, 6 × 8 = 48, 7 × 8 = 56, 8 × 8 = 64, 9 × 8 = 72" and has been told by her teacher to say each one 25 times and then to see if she can say them without looking at the paper. The girl practiced faithfully and could say them all when she went to bed but in school the next morning she was unsure of 7 × 8 and 9 × 8. Privately she then formed an association for remembering. For the answer 56, she thought of 5, 6 as a sequence preceding 7 and 8. For the answer 72, she was told by a classmate to think of 80 and 8 less or 72. Note that her first learning was characteristic of the rote drill of 25 years ago and that the results were uncertain. Note also that she discovered a way to remember 56. Only the answer 72 shows a result based upon some understanding and this was furnished by another pupil. Here it is worth noting that pupils may do peculiar things to help them to remember if not to learn. Her drill might have been more fruitful and more enjoyable had she had opportunity to discover, to think, and to reach conclusions with guidance from the teacher.

3. A group of junior high-school pupils is learning to estimate the size of an angle within the range 0° to 180°. Previously these pupils have developed concepts of angles and angle measurement. They have constructed angles with compasses and with the protractor and they have had experience measuring angles. These

previous learnings are not only worthwhile in themselves but the sequence is propaedeutic to learning to estimate the size of an angle. For learning to estimate, the teacher draws angles of several sizes on the board, pupils estimate, and the estimates are checked by measurement. A pupil suggests using a reference line (right angle) to assist in the judgment. Such reference lines are sketched. Other reference lines such as bisectors of 90° are imagined. The teacher provides each pupil with a sheet of paper having 20 different angles in various positions and asks them to write their best estimates of size. These estimates may later be checked with a protractor. If reasonable limits of estimate have not been achieved by some pupils, further instructional helps can be given and more practice can be provided. Note in the above that a meaningful approach was used and that the larger portion of practice followed an opportunity to learn. This practice was closely associated with learning and helped to provide sufficient thinking, understanding, and drill to "fix" a more lasting impression. Of course this teacher had used the procedure with previous groups and knew about what to expect, how much practice was needed, and when it was opportune.

The three illustrations were cited to show the meaning and significance of drill in learning mathematics. Drill can be of many types; it can be visual, manipulative, oral, written, or any combination of these. To be of most value it must always be accompanied with good mental processes. Later in this chapter, a brief discussion of opportunities and needs for drill in learning the two topics, fractions and equations, will be presented.

PSYCHOLOGY AND DRILL PROCEDURES

Psychology is concerned with modes of facilitating learning or, more currently, "change in the behavior" of the individual. The role of drill in learning is both recognized and respected. But this is a drill as previously described and not the abstract rote drill which for so many years characterized the teaching of mathematics. Investigations and the accumulated experience of experimentalists in education tend to agree on many principles in the psychology of drill. Those that are most applicable to learning mathematics are the following:

1. The educational climate, atmosphere, or rapport of a class has a tremendous effect upon learning.

2. The ideas associated with incentive, drive, purpose, and goal have a strong bearing upon learning.

3. Schools (pupils) achieve just about what is reasonably expected of them. That is, unless it is an exceptional case, a school in which the achievement in arithmetic is poor, has not honestly tried to achieve good results.

4. For many pupils and for certain types of situations the initial response or conclusion in learning seems to have a more lasting impression than subsequent responses.

5. Factors that are almost indiscernible frequently effect learning. These include community mores, status of the school, dress and whimsey of the teacher.

6. Pupils like to make progress and to know when they are learning. They respond to praise more than they do to condemnation.

7. Children should become organizers, systematizers, groupers, and classifiers of learning instead of "isolators" thereof.

8. At certain ages or occasions children seem to delight in rote learning particularly if there is a rhythmic cadence or a sing-song sound.

9. Children, particularly younger children, seem impelled to use the kinesthetic avenue of learning.

10. The mode or avenue through which a thing is learned seems to have an effect not only upon the enjoyment of learning but also upon the rates of both learning and forgetting.

PRINCIPLES OF DRILL

Over the years several principles concerning drill have become recognized as generally sound and applicable to many situations in the teaching of mathematics. It must be remembered however that conditions surrounding a circumstance or situation may be the critical factor, and thus a general principle is only relatively sound. The following have basis in experiment and in tested teaching:

1. The learner should both understand what he is practicing and appreciate its significance to him as an individual.

2. The learner should have sufficient propaedeutic experience so that the *newness* in what he is practicing does not create a mental block for him.

3. The learner should be an active participant both in setting his goals and in the thinking-striving aspects of learning. He should not merely repeat "parrot-fashion" from a teacher or textbook.

4. Drill should follow the developmental and discovery stages of learning and be used to reinforce and extend basic learning.

5. Drill should be varied so that procedures do not become monotonous and so that different pupils have types of drill perhaps better suited to them.

6. Drill should be spaced so that (a) time is not wasted in excessive overlearning in initial stages and (b) previous learnings are kept fresh and useful.

7. Drill should be an integral part of various phases of learning but should not be used to hasten the achievement of results at the sacrifice of meaning and understanding.

8. Drill policy should recognize different rates and modes of learning with different pupils and not try to fit all into a common mold.

9. In general, it is better to provide for drill upon whole processes rather than parts thereof, unless some particular part such as, for example, subtraction in a long division exercise causes trouble and needs teaching and practice for reinforcing.

10. Drill should be done with correct processes lest a child practice errors which need to be remedied later.

11. Drill should be based upon or involve thinking and insight so that it never becomes a mere mechanical repetition.

12. Drill should be used when and where needed. It should not be used as a punishment nor should things already well learned be assigned for more practice.

13. There should be some sense of organization of drill so that (a) pupils see the sense and relationships of what they are doing, and (b) important elements are not overlooked.

14. It seems that pupils of lower mental abilities require more

drill than the more able but this may be due to other related factors such as attention, insight and other such causes.

THE USES OF DRILL

What useful purposes are served by drill procedures in the teaching of mathematics? Why is it necessary to repeat a thought or a performance to insure learning? Is it not possible for pupils to gain complete learning by gaining insight and thus require none of the repetitive work? Under certain conditions, with certain pupils, and with certain materials complete and lasting learning seems to be achieved in one experience. All good teachers can testify to this but they also admit that in most cases drill procedures are required. The old adage "practice makes perfect" is unsound logically and untrue experimentally. In any learning situation there are many variable factors which a teacher can only partially control and hence it is not possible to prescribe precisely *when*, *where*, and *how* to use drill. However there is now general agreement that drill, practice, or recurring experience are useful:

1. To fix for more facile recall and for greater usefulness information whose significance is understood; for example, the fact that 60 minutes equal one hour.

2. To gain proficiency in handling a mathematical process or procedure after it has been studied and its usefulness established; for example, subtraction, solving equations and other such activities.

3. To enhance and enlarge the understanding of a concept whose basic principle or idea has been established; for example, drill upon subtraction situations or upon the concept of the tangent of an angle.

4. To improve the understanding of and the ability to use a generalization after it has been developed and stated; for example, the generalizations "cost equals number times price" and "quantities equal to the same value are equal."

5. To review and refresh processes after a period of disuse; for example, column addition, and solving linear equations.

6. To encourage and develop the ability and the will to speculate, to discover, and to discern in terms of mathematical relation-

ships and principles; for example, to discern that 6 × $4.98 is 12¢ less than $30.00, and to recognize linear relationship of two variables from a chart of values.

7. To learn to sense the mathematics in a situation and conversely to project mathematical principles into socio-economic situations and draw valid conclusions in terms thereof; for example, buying by dozen or pound or estimating the amount of meat needed for 20 people.

8. To gain confidence through mathematical success and thus erase the fear that many people have of a situation that is mathematical; for example, note how many times someone fears to record and total the bridge score.

9. To feel the thrill of achievement of having mastered something. This is different from but yet comparable to mastery in a physical task or sport.

EXAMPLES OF DRILL IN MATHEMATICS

Drill is a valuable part of learning. It should never be a sole mode of learning. It naturally follows the discovery and developmental phases of learning. Drill is conceived to be practice or recurring experience in which there is a recognizable element of similarity from one experience to another. This element of similarity must be apparent to the learner. In this discussion drill is not mere repetition. It is extending and applying previous experience for the sake of learning which is important to the individual learner. Drill procedures are applicable to all aspects of learning and should be employed. Thus drill applies to the learning of concepts, the development of mathematical principles, the mastery of a process, the ability to sense a problem situation, the reasoning through a situation, the feeling of a need to verify or check, and the final ability to use mathematics in the world of affairs. In addition to these mathematical uses, drill and experience ought to be used in the development of a spirit of inquiry and discovery, in fostering good habits of thinking and work, and the desire and ability to discern and judge.

It should be remembered that the methods of drill in learning mathematics cannot be reduced to simple laws governing the factors: (a) when to drill, (b) how much drill is needed, and (c) what form should the drill take. It is not wise to attempt to use laboratory results or piecemeal researches in setting a program of drill or of total learning. A human being is more than an assemblage of its parts and a public school has many influences and conditions that make "controlled learning" untenable as a postulate.

In order to show in brief scope the range of possibilities in drill, illustrations will be given for two topics: (a) fractions from the arithmetic of the elementary school, and (b) equations from algebra. In each case learning will be carried from the stage of concept development to functional usefulness in real life.

DRILL IN FRACTIONS

1. *The concept of fractions.* As children learn about *parts* as *fractions* of some whole thing such as an apple, cookie, or piece of paper, and the associated words of half, third, and fourth, they should have experience in actual cutting, in drawing lines to show, in talking about, and in recording or writing fractions. Each of these experiences is practice or drill, it is both recurring and developmental and is a vital part of learning. Similarly, the idea of a fraction representing the part or parts of a group or collection of items such as cows, people, and books must be presented, experienced, practiced. And other phases of the fraction concept such as the comparison of two groups and expressing this as a fraction must have drill at the appropriate time. Later when the fraction as an expression of comparison is extended into the concept of ratio, this too requires practice. At all stages it is desirable to "fix" ideas through recurring experience. This includes manipulative experience, visualization, oral mental, and written mental. Usually, concepts develop slowly and it is desirable to enlarge them over a long period. This applies particularly to those such as fractions in which there is a large variation in complexity.

2. *Principles of fractions.* It is probably unwise to set separate and distinct practice of most of the general principles of fractions because these are better handled as outgrowths and developments from the concepts of fractions and the need and use of them in such operations as addition and multiplication. But, these principles need to be developed and practiced. The most important are: (a) the whole of anything is the sum of its parts and such

attendant ideas as 4/4, or 5/5, or 8/8 make up one whole and when a part is lacking, the remainder (from a whole) can be easily determined; (b) the meaning and function of the two numbers that make up a fraction and the relationship of numerator to denominator; (c) the effect upon the value of a fraction when both terms are multiplied or divided by the same number and when the same number is added to or subtracted from both terms; and (d) the various relationships of common fractions to decimal fractions ranging from concepts to principles and manipulations. Principally drill or practice must be on the understanding and use aspects so that children learn to think with and in terms of the symbols and are able readily to answer such questions as "How much is it?" "Which is more?" "What happens if . . .?" It will be apparent that visual and manipulative impressions leading to principles will need practice as well as the principles represented in symbolic notation. Each of these is a reinforcement of the other.

3. *Computational skills with fractions.* Again, skills should rest upon a basis in understanding and this understanding comes through several avenues: (a) visual impressions, (b) manipulation of real things and models, and (c) study of the numerical facts and relationships in the symbols. Practice of each of these types of learning which lead to computations and manipulations with fractions is needed. The following types of abstract or process work also require practice if the pupil is to gain proficiency and independence: (a) changing from one fraction to an equivalent fraction (reduction, mixed numbers, improper fraction, and such) for use in comparisons, judgments, and computations; (b) adding and subtracting fractions; (c) multiplying and dividing fractions; (d) expressing a common fraction or ratio as a decimal fraction and in equivalent percentage notation; and (e) raising to powers and taking roots. In general whole operations of the simpler types should be learned and practiced first. However, special practice on a sub-step often is desirable as for example the case of "changing" in the minuend in subtraction. An examination of textbooks suggests that most drill in fractions is given to this group of manipulative skills. However, that should not be the case if genuine meaning and functional competence is held as the aim of instruction. Most good schools now have models and visual aids which

may be used by pupils to gain insight into operations with fractions. Further understanding and significance is achieved through study and practice with the relationships inherent in the symbols and notation of fractions.

4. *Functional competence with fractions.* Ability to (a) sense a use of fractions; (b) recognize and understand the essential principles involved; (c) know what to do, to think through the situation; (d) perform the necessary steps of computation; and (e) verify and feel confident of a conclusion are phases of a genuine functional competence with fractions. The development of these abilities is not an automatic consequence of study unless the work is directed to that end. Thus, as the various phases of fractions (concepts, principles, and skills) are being learned, they should be tied to functional situations. This tying to experience and in turn the study of experience for its fraction content are things that require drill or practice. Children tend to learn that which they try to learn. Certainly we cannot expect them to learn to sense uses of fractions and to use them if they are not given opportunity to practice this. And this practice must involve a good deal of thinking.

DRILL IN ALGEBRAIC EQUATIONS

1. *Readiness and pre-equation learning.* In all elementary schools much practice is given to number facts stated in equation form; e.g., $3 \times 7 = 21$ and $16 - 9 = 7$. This should and does provide a basic understanding for algebraic equations which come much later. Similarly, the use of the question mark (?) to indicate an unknown value is common practice; e.g., $5 + ? = 11$, $7 \times ? = 56$, and $\frac{3}{4} = ?/12$. At a later stage, the statement, the writing, and the evaluation of formulas provide a basis for equations; e.g., $I = PRT$, $A = \frac{1}{2} bh$, and $C = \pi d$. The thinking involved in solving inverse cases based upon formulas is essentially algebraic. A number of elementary school experiences lead directly to basic axioms of algebra; for example, the equality of 2 dimes and a nickel, a quarter, and 25 pennies. Thus it is apparent that both in modes of thinking and in technique the work of Grades I–VII has laid a basis for equations. However, the value of this basis depends in large measure upon the methods of thinking and of

work used by teachers and pupils. All of these items have been involved in recurring experience or drill. Another item of learning that is important in writing equations and in thinking about them and which is usually given separate practice is that of writing algebraic representations for unknown quantities and relationships involving them. For example, the age of George 9 years ago may be represented by $X - 9$ if X is his present age, or by X if his age now is $X + 9$. This aspect of equation writing is so important it should receive considerable practice.

2. *Solving linear equations.* The solution of equations follows a fairly well established pattern in terms of sequence with solutions based upon basic axioms coming first and followed by short-cut methods employing such techniques as "transposition" and "multiply by the common denominator." As in other work, it is most desirable for basic learnings to precede practice so that drill may be meaningful and fruitful. The following steps or phases of solving linear equations require practice: (a) solutions employing separately and uniquely each of the equality axioms of addition, subtraction, multiplication, and division; (b) solutions using the basic axioms in combination; (c) solutions using short cuts such as transposition and "cross-multiply"; and (d) verifying and checking solutions. Most textbooks provide adequate amounts of practice in the solution of equations. They do not, however, and probably cannot, direct the thinking of the pupils so that they reach a high level of appreciation and understanding. This understanding tends to result in a slower rate of forgetting and also it provides something to rebuild upon when a pupil is temporarily stymied in a solution. Furthermore it provides the basis for transfer from one type of exercise to one that is slightly different.

3. *Writing equations.* The real essence of algebra is probably more involved in the writing of algebraic relationships and equations than it is in the manipulations involved in the solution of equations. The ability to represent algebraic relationships in equations is rather subtle; it is not one that can easily be isolated and drilled. It involves a collection of abilities which seem to be enhanced by native intelligence. However, drill, practice, or experience upon certain phases of algebraic representation and with

particular emphasis upon thinking seem to be fruitful. These are: (a) appreciation of the significance of and the ability to locate the more basic unknown value and represent it symbolically as for example by X or N; (b) learning to think and to express orally and in writing simple relationships; e.g., $2X + 5$ represents five more than twice some unknown value; (c) the ability to think through and to represent the combination of algebraic expressions by processes such as addition and also to express equality in an equation; (d) ability to rearrange or reverse the relationships leading to an equation and thus provide a second approach or check to a solution of a basic problem; and (e) ability to write equations from any reasonable situation or problem where relationships are linear.

SUMMARY

1. The terms "drill," "practice," and "recurring experience" are used to denote that aspect of learning which has a recognizable element of commonality that is repeated. While these terms are not synonymous, they suggest a broader vision of the nature and role of drill than that commonly held a decade ago.

2. Drill, broadly conceived, is both important and necessary in learning mathematics. It is really a part of the learning process and when properly applied aids in understanding as well as in proficiency. The importance and need for drill are little conditioned by the "brand" of psychology one accepts.

3. Because of the many attendant and variable factors in a classroom it is not possible to state precise rules governing the *when*, the *how much*, the *where*, and the *how to* use drill.

4. Drill, practice, or recurring experience should be used with all phases of learning mathematics beginning with methods of discovery, the development of concepts, habits of work, and carrying through to the computations, thinking, and judgment that are essential in achieving functional competence and independence for the individual.

5. It is the teacher's responsibility to plan appropriate experiences to provide, at optimum times, for learning and drill.

6. The diagram below shows elements and relationships in

learning mathematics. Recurring practice or drill is needed not only on the several aspects or elements but also on the connections between them.

Mathematics
Concepts – Information – Principles – Relationships – Solutions
Experience → Learning ———→ Experience
↕ ↕ ↕ ↕ ↕ ↕
Incidence → Discovery → Understanding → Functional Use

7. Transfer of Training

Myron F. Rosskopf

One of the leading students of transfer of training summarizes in a recent study the results of experimental research in the following words:

First, transfer is a fact, as revealed by nearly eighty percent of the studies; second, transfer is not an automatic process that can be taken for granted, but it is to be worked for . . . ; and third, the amount of transfer is conditioned by many factors, among which are: age; mental ability; (possibly) time interval between learning and transfer; degree of stability attained by the learned pattern; "knowledge of directions, favorable attitude toward the learning situation, and efficient use of past experience"; accuracy of learning; "conscious acceptance by the learner of methods, procedures, principles, sentiments, and ideals"; meaningfulness of the learning situation; the personality of the subject—greater transfer in extroverts than in introverts; method of study; suitable organization of subject matter presentation; and provision for continuous reconstruction of experience (12).

All of these factors that condition transfer are relevant, but only the last two will be considered in any detail in this chapter.

Transfer of training theories change as psychological theories of learning change. Each new development in the psychology of learning leads to new experiments on transfer of training and to reinterpretation of the results of past experiments. In order to understand currently accepted conclusions with respect to transfer of training, it is necessary to know the background for these conclusions.

DOCTRINE OF FORMAL DISCIPLINE

The doctrine of formal discipline is based on what is known as "faculty psychology." Faculty psychology postulates that the mind is composed of several faculties such as the will, memory, judgment, and the like. The theory of learning called formal discipline holds that these mind faculties can be trained by exercise. The material studied or learned is not important but the hard work for the mind involved in the study and in the learning is most important.

The point of view represented by the Committee of Ten of the National Education Association is typical of the theory of learning practiced in the latter part of the nineteenth century. The Committee of Ten was organized in 1892 to survey secondary-school practices and to make recommendations for improvement of practices; the following sentences reflect the point of view of the majority report:

> The mind is chiefly developed in three ways: by cultivating the powers of discriminating observation; by strengthening the logical faculty of following an argument from point to point; and by improving the process of comparison, that is, judgment ... studies in ... mathematics are the traditional training of the reasoning faculties ... (11).

Thus, it was held that formal work is the best way to facilitate transfer of judgment, reasoning, and observation to problems of living. The particular school subjects studied do not make much difference, it was believed, so long as they are difficult (provide sufficiently hard exercise for the faculties of the mind) and can be presented in a series of formal lessons. The improvement in reasoning acquired in mathematics, for example, would so develop the faculty of logical thinking that there would be transfer of the ability to reason logically to history or science or languages.

The doctrine of formal discipline is, of course, discredited as a means of learning that facilitates transfer. Even some members of the Committee of Ten disagreed with the majority report and presented a report of their own. Some nineteenth-century psychologists realized that there seemed to be little carry over from training received in one area to another area. For example, William James experimented with improvement of memory and found results based on formal discipline to be very unsatisfactory. The death-blow was struck by the publication in 1901 of the thorough and scientific investigations of Thorndike and Woodworth. From that date, no psychologist insisted upon faculty psychology or formal discipline. And yet, methods of instruction that are used by many teachers today are based on formal discipline. The appeals of lay people (and some educators) for a return to the good old days of really "hard" instruction imply a theory of learning based on formal discipline. It is distressing to find so much mathematics and so much mathematics teaching based on an outmoded theory of learning. Authoritarianism is seen so often in a mathematics classroom; a teacher points out the correct response, students accept it, and then practice applications of the correct response. The practice exercises are graded from simple to difficult in "good" textbooks, but there is little opportunity provided for students to explore or to discover or to organize experiences. During the past 50 years much sound experimental evidence has been accumulated that proves such instruction to be ineffective in the promotion of transfer.

Instruction in upper secondary-school mathematics particularly, easily falls into the stereotyped form implied by the doctrine of formal discipline. Because of the rigor required by logically organized subject matter, it seems difficult to break away from teaching that is showing students how to reach correct solutions to problems. However, there are accounts of experimental work in mathematics classrooms (9, 5) that are encouraging. Teachers in these classrooms are attempting to put into practice a theory of learning that is currently acceptable and, that experimental evidence indicates, promotes maximum transfer.

DOCTRINE OF IDENTICAL ELEMENTS

As was pointed out in a foregoing paragraph, many psychologists and educators protested the doctrine of formal discipline on philosophical and logical grounds. William James is credited with being one of the first psychologists to test the doctrine of formal discipline experimentally. His experiments, conducted about 1890, are crude according to contemporary standards of psychological research, but the results that he obtained showed that formal discipline has little effect on improvement of memory. The account of the historic experiments of Thorndike and Woodworth were published in 1901. It is in these papers that the doctrine of identical elements was stated: "Spread of practice occurs only where identical elements are concerned in the influencing and influenced function" (15). Their method was to give students practice in estimating the areas of rectangles varying in size from 10 sq.

cm. to 100 sq. cm. They found that the students showed considerable improvement in estimating the areas of small rectangles if they were given the correct area after each estimate. But when the students were presented with the problem of estimating the area of a large rectangle or the area of a figure of a different shape, it was found that the students showed little improvement.

As experimental evidence accumulated during the next 20 years, Thorndike and others added to the theory of associationism. For it is from the psychological theories of learning called associationism and connectionism that the doctrine of identical elements emerged. There is a question of what is meant by identical elements. Are identical elements to be understood in terms of training specific individual abilities? Or, are we to understand that identical elements in two situations include both specific abilities and the statement in words of a principle used in a learning situation? Woodworth in a recent statement believes that a correct formulation of the theory is as follows:

> The more definitely the principle is isolated, even to the extent of formulating it in words, the more chance of transfer ... if the principles are embodied in words, they are concrete bits of behavior and their transfer from one situation to another creates no difficulty for the theory of identical elements (18).

According to Gates' interpretation of Thorndike's theories, he "used as equivalent to 'elements' such words as 'aspects,' 'factors,' 'features,' and 'relations'. . . . His concept . . . can, in fact, include anything as 'elements' which investigation proves to be actually operative" (3).

Hence, it appears that Thorndike developed and expanded his theory as evidence accumulated. Early interpretations of identical elements as specific components of a learning situation were extended to include words or components that were complex. According to an eclectic point of view, the interpretation of "identical elements" can be as broad as the interpretation of "structure" by gestaltists. For example, Thorndike writes,

> The newer pedagogy of arithmetic, then, scrutinizes every element of knowledge, every connection made in the mind of the learner, so as to choose those which provide the most instructive experiences, those which

TRANSFER OF TRAINING 209

will grow together into an orderly, rational system of thinking about numbers and quantitative facts (14).

Later in the same volume when he is discussing the psychology of drill in arithmetic, he says:

As each new ability is acquired, then, we seek to have it take its place as an improvement of a thinking being, as a co-operative member of a total organization, as a soldier fighting together with others, as an element in an educated personality. Such an organization of bonds will not form itself any more than any one bond will create itself. If the elements of arithmetical ability are to act together as a total organized unified force, they must be made to act together in the course of learning. What we wish to have work together we must put together and give practice in teamwork.

... *every bond formed should be formed with due consideration of every other bond that has been or will be formed; every ability should be practiced in the most effective possible relations with other abilities* (14).

The phrase "every bond formed" in the foregoing quotation might be interpreted to mean that an operation in arithmetic, or mathematics, is to be learned through direct practice of the operation; that is, an operation becomes "fixed" in the mind of a student through repetitive doing of exercises involving the operation. An extention of this interpretation would be that transfer is achieved through drill. Associationists say that this is a narrow view of the doctrine of identical elements, that bonds *are* formed through direct practice, but that the direct practice must be of such a kind that it takes into account other (related) bonds, together with the attitudes, fatigue, set, purpose, and the like, of the learner.

One other observation will be made concerning the doctrine of identical elements. The percentage of transfer from one learning situation to another learning situation is always less than 100 per cent. Direct practice in one learning situation increases the success in that situation but the success in another learning situation is proportional to the number of identical elements in the two situations. It seems, then, that the amount of transfer will depend upon doing over again in a second situation those elements or components that are common to it and a first learning situation.

Orata in his most recent survey of the evidence for transfer of training states, "As the theory of identical elements tends to become obsolete, the role of insight and generalization becomes more thoroughly established (12)." Orata has made an intensive study of the doctrine of identical elements and its relation to transfer of training. In the experimental studies that purport to support the theory, he clearly shows weaknesses both in experimental design and interpretation of results. Because the drill theory of instruction is based on the doctrine of identical elements, it is necessary to discuss it in some detail. More and more evidence is accumulating to indicate that students do not learn, in any sense of being able to transfer or to apply what they have learned, by practicing processes in isolation.

Practice of the same response merely increases facility in producing that response, whatever its nature and its level of usefulness and maturity. If one repeats the definition of some term without understanding its meaning, one cannot through repetition acquire meaning for the term, however proficient one may become in saying or writing or thinking the definition.... For the definition ... to possess meaning, the learner must respond to the definition ... in a variety of ways (2).

However, it may not be wise to discard the doctrine of identical elements entirely when we think in terms of educational practices. There is an interpretation that can be given to some experimental evidence that indicates there are important applications of the doctrine. It is a question of timing and place. There are different levels of work in mathematics. At one level a student is exploring and discovering relationships (organizing his experience) in order to arrive at a principle that can be used to solve new problems. This is true of arithmetic, for example, in the introduction of each new process. But when a student has passed this level and gone on to a new process, his response to the old process must be almost automatic if he is to master in any effective way the new process. When one comes in his school experience to the process of multiplication, he must have passed beyond the exploration stage in his mastery of the process of addition. Similarly, in the teaching of trigonometry, the fundamental identities are developed with a class by helping them to organize their experiences with the trigonometric functions and to understand the relationships that exist among them. Students must go beyond this understanding of the organization of relationships among the trigonometric functions if they are to be successful in proving trigonometric identities. They must be able to give an immediate response to any trigonometric function in terms of its related identity. With this mastery, attention can be given to exploring on a wider frontier. Without this learning or mastery, a student will find the proof of trigonometric identities a hard, laborious task.

It is exactly in situations of the foregoing sort that drill, based on the doctrine of identical elements, has a place in the teaching of mathematics. That this is true is supported by the following experiment carried on by Katona. Katona had three matched groups. All groups were taught by a meaningful method but Group I was taught three tasks, tested immediately, and then four weeks later; Group II was taught the same three initial tasks but in later teaching periods was taught different tasks that involved the same principle as the initial tasks; Group III was taught the same three initial tasks and in later teaching periods practiced these tasks. All groups were tested at the end of four weeks on the same three tasks as were taught initially, as well as on new tasks. In describing the performance of Group III, Katona writes,

> By reviewing the performance of this group solely with the practiced tasks we find a perfect example for a practice curve. In the first intermediate test a slightly lower score was obtained than in the immediate test, but from the first to the fourth week the improvement proceeds in a straight line. In the fifth test (main test) the 21 subjects of the group committed only one error in solving the three old tasks.... Observation of the behavior of the group shows that we have here a performance strictly comparable to the well-known effects of practice by drill.... There was no "solving" of a problem, but rather a recall of well-learned data. Here we find reproduction instead of reconstruction (8).

In a footnote he quotes Breslich as follows: "Simplified routinized processes are sometimes the outcomes of earlier understandings that have been reduced to formulas which are merely held in memory" (8).

The point is that some processes in mathematics are used so much in subsequent work that a student must have the sort of mastery that is of the stimulus-response sort. That is, a student must have this sort of mastery if he is going on into subsequent mathematical work. In order to understand long division, one must be able to substract and multiply; if a student is hesitant in his multiplications and subtractions, this hesitancy is going to get in the way of his learning division. For learning elementary algebra, a student must be able to perform arithmetic operations with facility. One could give example after example of the sort of mastery that is described here. A teacher must make a distinction between learning understood processes by drill methods and helping to organize experiences in the development of new concepts in order to achieve understanding. This is what was meant by an earlier sentence stating that the doctrine of identical elements depended upon timing and place for its use.

DOCTRINE OF GENERALIZATION AND MATURATION

Judd was one of the first psychologists in America to differ with the doctrine of identical elements. In his writings he emphasized the importance of generalizations. He experimented with subjects who were taught the principle or generalization involved in a task and compared their performance with subjects who did not know the principle. His objective was to test the effect on transfer of knowledge of a generalization. Although his test group was small and there was no control group, yet his conclusion that knowledge of a principle facilitates transfer had a great effect (7). Other psychologists with better experimental techniques did research that tended to support Judd's conclusions.

Beginning about 1912, a group of German psychologists published papers in which they criticized the attempts of association psychology to reduce the study of mental activity to elementary and individual connections. These psychologists, Wertheimer, Koffka, and Köhler, emphasized the importance of considering complex "wholes" in order to understand mental activity. This school of psychologists became known as gestaltists because of their insistence that a human being reacts to a whole situation or structure, rather than to the individual parts of a situation.

They asserted that a person achieved undertstanding of the parts of a structure only through an understanding of the whole structure. Out of elaborations of this thesis grew the familiar statement of gestaltists that the whole is greater than the sum of its parts.

Since much of Judd's later writings uses some of the language of the gestaltists but continues to stress the importance of generalization, some quotations follow that are illustrative of his thesis.

It is of importance for an understanding of the nature of the higher mental processes that there be clear realization of the fact that it requires time and laborious reorganization of experience for the individual to gain full comprehension of the meaning of the words which make up the number system (6).

Notice the emphasis upon "reorganization of experience" of the individual. This phrase might have been made by a gestaltist, for the gestaltists believe that insight or understanding or generalization comes to an individual through reorganization or reconstruction of experiences with a whole situation. It is this reorganization of experience that requires time for maturation. For example, spaced drill has been found to be more effective, so far as transfer is concerned, than concentrated repetitive drill.

At the higher levels of arithmetical thought and manipulation as well as at the lower levels, it is not enough that the mind acquire mere rules or successions of isolated ideas. There must be an organization of experience of a form that is . . . described by the term "conceptual."

The view that all mental activities can be explained in terms of elements which are of the simplest and most primitive type overlooks altogether the principle that . . . organization . . . accounts for life.

If psychology is to rescue education from the new formalism, which consists in devotion to mere acquisitions of detached and unorganized facts—if mathematics, the natural sciences, and all other school subjects are to be taught by some method other than mere drill—there will have to be clear recognition of the difference between the lower and the higher forms of mental activity. The higher forms of experience will have to be emphasized as the true ends to be reached by the processes of education (6).

Snoddy (13) performed an experiment with human beings in which the task consisted of drawing a star pattern by looking in a mirror. He found that the movements of a subject were jerky and undifferentiated at first. When a subject came to a "corner" there was a delay in his movements; then, when insight into the process occurred, movements speeded up and became smooth. He called this hesitation a "period of initial delay." Time had to be allowed for an individual to organize his perceptions, to differentiate among elements of the pattern; in short, the individual needed to see how to proceed to make the "corner."

In writing about the teaching of arithmetic, Wheeler and Perkins say:

A given number derives its meaning from its position in a whole. 4 is not only 4 but so much more than 3, so much more than 2, so much less than 5, so much less than 6. Just how much the difference is, must be discovered in the course of maturation, induced through the stimulation attending the use of the numbers (17).

It is through many experiences with numbers or, more generally, mathematics that a student grows in maturity of organization of his experiences. Only through many experiences can a student achieve the differentiations that lead to an understanding of the meaning of numbers.

The grouping method of teaching number facilitates the evolution of relationships in the child's thinking... the teacher cannot adequately present the subject [that is, arithmetic] who does not understand the logic of number herself. She must give to her pupils the simplest little problems that will bring out the early forms of configurational response to number stimuli (17).

A discussion of the doctrine of generalization and maturation is important, because of its wide use in the teaching of mathematics. Typically in secondary-school mathematics classes, a generalization or principle is presented by a teacher, illustrated by applications, and then students are assigned problems that require use of the principle in their solution. Such classroom instruction depends upon students, understanding how a principle is built up step by step from more elementary generalizations. In this respect the instruction is good. Certainly, it is much better than instruction that consists of presenting a way of doing the type of operation under consideration and assigning practice exercises. Early in his investigations Katona observed a difference in behavior of subjects taught by a method to promote understanding. These observations led to experimentation with methods of teaching for understanding. He compared the performance of a group taught by the method of presenting the principle or generalization needed to solve a certain task with the performance of a group taught by another method. At the end of a lapse of four weeks it was found that the group taught by the method of generalization performed better on old and new tasks than a control group or a group taught by memorization. None of the groups performed as well as a group taught by a method that stressed the structure of the task. In a footnote he writes about the group that was taught by the generalization method as follows:

Sharp improvements and unexpected deteriorations in the accomplishment of individual subjects were the rule rather than the exception. Learning by the abstract principle is thus characterized in our case by a certain degree of instability (8).

DOCTRINE OF REORGANIZATION OF EXPERIENCE

McConnell writes, "Newer trends in the psychology of learning emphasize the primacy of organization" (10). It is easier to illustrate this principle than it is to describe the principle. Human beings, as well as animals, react to a constellation of stimuli rather than to each element-stimulus of the constellation. It is the structure or the organization or the relationships that a student sees in a mathematics problem that permits him to arrive at a solution. For example, Brownell writes:

What one does when one learns is to attack the new problem with whatever reactions are available. These reactions are seldom if ever of the blind trial-and-error variety, but represent forms of behavior which have been connected previously with some aspect of the problem situation. Thus, at the very outset of learning, we encounter organization of behavior . . . (1).

On the basis of experimental studies, Brownell concludes that learning takes place as a result of reconstruction or reorganization of experience. Trial-and-error learning in the usual sense of the phrase does not exist. A student may seem to be making efforts at random to solve a problem as an observer who knows how to solve the problem looks on. But these efforts are not at random at all for the student. They are evidence of his attempts to organize his past experiences, to reconstruct from them a way of attack, to see relationships between this new problem and past problems.

Katona has done extensive experimental work on learning by reorganization of experience. He makes the following comments concerning the results of a card trick problem that he presented to students.

> These observations . . . characterize the process of reproduction [of the card trick] on the part of the "meaningful learners." The subjects proceeded to discover or to construct the solution, and the preceding training helped them to do so. Reproduction was not at all similar to a door bursting open, because a button has been pressed—it did not consist of the presentation of an ever-ready response to the appropriate stimulus. It was more like the processes of discovery, of problem-solving, and of construction. Remembering can here be best characterized as a rediscovery—a *reconstruction*. The effect of learning was ability to reconstruct (8).

In summarizing the results of his investigation of methods of instruction to be used in his experiments on learning and on transfer of training, he makes the point:

> Both problem-solving and meaningful learning consist primarily of changing, or organizing the material. The role of organization is to establish or to discover or to understand an intrinsic relationship. . . learning by understanding consists of grouping (organizing) a material so as to make an inner relationship apparent (8).

Using a method of instruction that helped subjects to see the structure of a task and the relationships that existed between the form of the task and its solution, Katona found more than 100 per cent transfer. That is, subjects taught by this method performed the practice tasks about as well when tested at the end of four weeks as they did at the beginning of the experiment, but they performed better on new tasks than on the practice tasks. In his discussion of the experiment he writes:

> The process of gradual organization, the slow transition from a worse to a better state of affairs, from a bad to a good gestalt, is just as important for the-psychology of meaningful learning as is the flash of insight.
> We need the concept of gradual meaningful learning to understand the learning process.... The thesis that a single exposure is sufficient, that repetition is not required, cannot be justly applied to learning by examples or help. Only if we falsely define "repetition" as the repeated occurrence of the identical contents A, B, C, D and their apperception in an unchanged form at several successive presentations, may we say that repetition cannot occur in meaningful learning. But by using the term in a different sense, the successive steps in the method [of instruction] of examples may be said to constitute repetitions....
> Repetition ... is not repetition of one set of identical elements, rather it is a gradual development of structural features. One does not do the same thing over and over. On the contrary, one is always passing on to a more advanced performance (8).

To secure maximum transfer, in the sense of applying "an integrated knowledge, [a whole principle] ... to all tasks involving the same principle" (8), teachers of mathematics must teach in such a way that demonstration exercises (or tasks) serve as examples of the application of the principle. If the learning is directed by a teacher toward an understanding of how a well envisaged structural situation can be solved, a student's probability of success in applying the principle to a strange, different structure that requires for its solution application of the same principle will be greater than if learning is directed toward memorization or generalization. The principle need not be verbalized by students. In one of his experiments, Katona asked his subjects the following question:

> Try to formulate in a few words the main point or principle of the tasks on which you have just worked. What is the essential thing you

have to know in order to be able to solve such tasks? I do not need exact definitions; a hint at certain ideas you have in mind will suffice. You will have three minutes in which to write (8).

He found that the answers to this question were unsatisfactory.

There was no correlation between the few satisfactory answers and the performance of the same subjects as revealed by their scores. Many subjects who solved all or most of the tasks were, nevertheless, unable to give an account of the problem's main points.

This result reveals that it is very difficult to express in words what is required to solve such tasks. The ability to solve the tasks can be acquired without verbal formulation of what has been learned and successfully performed. Conversely, formulation alone is no guarantee of good performance (8).

Although Gertrude Hendrix frankly admits that her results need to be tested further, enough experimental work has been done so that the following hypotheses are emerging from the data:

1. For generation of transfer power, the unverbalized awareness method of learning a generalization is better than a method in which an authoritative statement of the generalization comes first.
2. Verbalizing a generalization immediately after discovery does not increase transfer power.
3. Verbalizing a generalization immediately after discovery may actually decrease transfer power (4).

The implications of the observations of Katona and Hendrix for the teaching of mathematics are clear. In the first place, when a class is introduced to a new concept there must be active student participation in discovering the concept and how to apply it. Such active student participation will depend upon adroit questioning on the part of the teacher and upon his sensitivity to the progress of the class in its exploration of the concept. As soon as the students are *aware* of the concept, they are ready to apply the concept. They are ready for practice problems, if you like, or for an opportunity to solve new tasks that are different from the demonstration examples but require the concept (used here as a synonym for principle or generalization) for their solution. At this stage of progress of the students a teacher must be satisfied with students understanding how to apply the concept; no attempt should be made to have the students state the principle in words nor should the teacher give the statement of the principle in words. One might say at this stage that students should work intuitively. Stress should be put on "*grouping, reorganization, structurization,* operations of dividing into sub-wholes and still seeing these sub-wholes together, with clear reference to the whole figure and in view of the specific problem at issue" (16).

If achievement of maximum transfer is an objective of the teaching of mathematics, then at every level an effort must be made to use a developmental approach in a classroom. By teaching so that students reorganize their experiences and become aware of how the overall structure of a problem is related to its elements a teacher can achieve with those students a disposition to use mathematics that cannot be achieved as well by any other method. Every teacher believes that ability to state a principle in words represents a higher level of understanding. That is true, but this higher level of understanding, represented by ability to verbalize, is a level that is approached when a principle (or principles) is needed for investigation of a topic in mathematics. It is necessary that a teacher be able to verbalize principles and generalizations and concepts of mathematics if he is to be able to use the developmental approach in a classroom. A teacher must be conscious verbally of a principle in mathematics, even though it is not necessary for students at the beginning to go beyond being aware of the principle.

SUMMARY

Thus, we see that, except for the first, each of the theories of transfer of training discussed in this chapter has implications for the teaching of mathematics. It is not necessary to regard the theories as mutually exclusive; aspects of the doctrines of identical elements, generalization, and reorganization of experience are applicable in mathematics classrooms. Of the four theories of transfer that have been formulated, that of formal discipline is the only one that is thoroughly discredited. All of the others are accepted, if not totally then in part, by all groups of psychologists. There remains much experimental work to be done on transfer.

Not only do we need to learn more about what is transferred, but we need to experiment to see how transfer can be facilitated. But, and this is important for all teachers, experimental research indicates that transfer is a fact. How to make the percentage of transfer larger is a problem that every teacher recognizes and that every teacher works on in his own classrooms.

After Thorndike and Woodworth, the most important formulation of a theory of transfer of training is that given by Judd in his doctrine of generalization. Judd was the first to experiment with the effect of knowledge of a principle involved in a task on transfer. He pointed out the importance of understanding of concepts and experimented with the effect of such understanding on skills. In many respects, Katona's experiments are elaboration of this idea and a filling in of details. Teaching for understanding and teaching meaningfully are phrases that are common today. Katona and Hendrix give valuable hints on methods of instruction that promote transfer. Both stress the importance of discovery, of exploration, of reconstruction or reorganization of experience. Both stress the importance of non-verbalized knowledge of a principle. In the teaching of mathematics there has been too much insistence upon students' telling (verbalizing) a principle or a generalization and not enough observation of students' applying a principle. More attention should be paid to a student's saying he understands how to do a problem but cannot tell how to do it. For the percentage of transfer is larger when a meaningful method of instruction is used that does not stress verbalization of the principle involved in the assigned tasks.

A program of mathematics teaching that will develop the largest possible transfer might be outlined as follows: (a) Teaching should be for understanding; for developing concepts. This means that the methods of exploration, discovery, and organization should be used. At this stage a teacher should be satisfied with a student being able to solve tasks that require use of the concept for their solution; at this stage there should be no attempt made to have students or the teacher verbalize the concept (of course, it is not implied that verbalization by a particular student should be discouraged). By presentation of examples and working them out together, teacher and students can achieve the sort of understanding that seems to give maximum transfer. (b) After understanding is assured, enough practice is furnished students so that they will have an opportunity to reorganize or reconstruct experiences in terms of the concept involved. In case the concept is one that is a routine part of larger problems—like addition or multiplication in arithmetic, or operations with signed numbers in algebra—the practice should be of the stimulus-response type. In such a case drill has a definite place in a program of mathematics teaching. (c) Those students who progress to higher levels of mathematics study should learn to verbalize principles that are appropriate to their level of progress. For these students, a teacher should insist upon their being able to tell how they do problems. These are the students who are studying mathematics because they are going to use its principles in other areas or because they have a love of the subject itself.

From the accounts of experimental work that has been done with mathematics groups it is not clear what the implications of research on learning and transfer of training are for general mathematics courses or mathematics courses in general education. It is safe to conclude, however, that teaching for understanding and for formation of concepts should be paramount. But answers to the following questions await further research. How much practice work should a general education course in mathematics include? How much effort should be made in such a course to achieve an immediate response sort of learning? How much verbalization of principles should there be?

Discovery and exploration through many examples that use the same concept should be the means of instruction in a general education mathematics course and the end should be applications of the non-verbalized concepts to new problems. If this objective were achieved in a general education course, there would be more adults with a greater disposition to use mathematics in quantitative situations. A variation of this suggested conduct of a general education course could be teaching by discovery and exploration, presentation of the principle verbally, and use of the principle in examples but with no insistence upon memorizing the principle.

EXAMPLES OF TEACHING FOR TRANSFER

The following examples of evidence of transfer of training or of good teaching for transfer are not theoretical examples. These are classroom experiences of the writer or of others who have been kind enough to contribute examples.

At one time there were in a twelfth-grade mathematics class a few boys and girls who were outstanding students in painting but somewhat casual students in mathematics. Required of all students in the class was a special project that could consist of the construction of a model, a report on some library research, or a report on an extension of an individual's knowledge of mathematics beyond the requirements of the course. The mathematics department of the school had bought earlier two books on dynamic symmetry. It was suggested to the boys and girls interested primarily in painting that they read these books and do paintings laid out according to the tenets of dynamic symmetry. Two of the students became deeply interested in the problem, studied the available books thoroughly, and interested their whole class in painting in root squares and whirling squares. These two students each completed excellent paintings as their projects for the course. The paintings were abstract, of course, using dynamic symmetry for their design and mathematical figures for their material.

Because these students saw a relationship between their primary interest and mathematics, they began to be more serious students of mathematics. Both of their areas of interest benefited. The students' enthusiasm for dynamic symmetry caught the attention of their art instructor. She, too, began to study and to realize that geometry plays a role in painting. The total effect was most beneficial. Many other students in art studied mathematics because of the applications that were possible. Many students of mathematics studied art because they saw it as a field that might provide them with a lasting cultural interest. Both areas of work learned a new respect for one another's efforts.

Many writers believe that mathematics students should use a library more than they do. In order to stimulate use of the library by mathematics students, each student in each grade was required to read at least one book per semester that was related to mathematics. Since this requirement was soon followed by a similar requirement by other departments besides English and social studies—where normally much library research is done, some plan had to be devised to prevent overloading students with work. After discussion of several proposed plans, it was agreed that the English department would prepare booklists for each grade level that would incorporate the recommendations of all departments. Students could choose books from these lists. In order to secure a proper balance of reading of various sorts of books, the English teachers would urge their students to select books from many areas. A student could use the same book, for example one dealing with mathematics, for both his English class and his mathematics class.

In addition, the English department suggested that it would include in its spelling lists technical terms from all areas, as well as the names of the instructors. Every geometry teacher knows how difficult it is to secure correct spelling of words like "vertical" and "angle" in plane geometry. Mathematics teachers enthusiastically indorsed the suggestion of the English teachers. On the score of correct spelling of teachers' names, it can be said that it is disconcerting to find that a student has been a member of one's class for a whole year and still cannot spell correctly his teacher's name.

Teachers observed that students had a better understanding of the relationships that existed between areas of study. There was more insight into the social setting of the development of mathematics and science. Students came to understand that literature, politics, mathematics, and science were related to a total pattern of activity in a period of time. The history of mathematics became part of the understanding of living. With the additional help of the English teachers on spelling, not so much time had to be spent on this aspect of mathematics teaching. As more and more of the faculty were drawn into this work, there grew in the school an insight into the problems of reading and spelling that before had been in possession of the English teachers only. Both teachers and students benefited from the booklists that were arrived at cooperatively.

In order to attack an objective of education, sometimes called the development of the disposition to use reflective thinking, another school faculty proceeded as follows: Under the leadership of the head of the science department a group was formed that consisted of two mathematics teachers, a social studies teacher, an English teacher, and a language teacher. This group met to discuss their readings, the implications for classroom work, and their classroom successes or failures. From the many outlines of the steps in reflective thinking that had been published, the committee worked out a synthesis and an elaboration of the steps. The elaboration of the steps in reflective thinking was intended to give teachers some idea of the concepts to be developed. Many faculty meetings were devoted to a discussion of the committee's work. There was enthusiastic support of its findings from some teachers; others were indifferent, because they did not see a direct connection between their teaching and reflective thinking,

On the junior high-school level it was decided to present opportunities for students to exercise reflective thinking. Teachers would guide the learning process in such a way that the steps in reflective thinking could be followed, but there would be no effort made to make students conscious of the method. In mathematics, science, and social studies classes, there was observed outstanding success of the program. English and language teachers reported that there was little opportunity in their classes for presenting good problem situations. But as teachers' experience grew it was found that they became more sensitive to the possibilities for reflective thinking in their areas and, hence, provided more opportunity for students to develop their ability to think clearly. The concerted effort made by the faculty to teach for this school-wide objective of education led to observable increase in students' attention to consideration of proof on their level.

For senior high-school students, the steps in reflective thinking were made explicit. Because of their greater maturity, these students considered much more complex problems than the junior high-school students. However, there were greater handicaps to overcome in that the courses students were taking tended to be more formal. It was more difficult for teachers of senior high-school students to learn a new method of teaching, to present subject matter material in such a way that a problem situation existed in the classroom. The better work was done in science classes, with mathematics classes following along as best they could. It was observed that during the years when there was most cooperation in the teaching for reflective thinking (war service scattered the original faculty and the necessities of war activities required the attention of the remainder), there were more science students engaged in original investigations and many of these used their knowledge of mathematics to help develop their science projects.

The opinion of the faculty was that a disposition to use reflective thinking could be fostered in students. Individual teachers contributed anecdote after anecdote to support this point of view. But, it was also the consensus that teachers needed to hold their attention firmly on a major objective like reflective thinking if the proper method of teaching for it were to be used in a classroom. The particular emphasis given to one or another phase of reflective thinking did not make so much difference as the method of teaching used. At every opportunity teachers had to be careful to provide an opportunity for students to explore a problem and to avoid explaining the generalization that would solve the problem.

One of the most clear-cut examples of transfer was contributed by Helen M. Walker, Professor of Education, Teachers College, Columbia University. The anecdote is the result of a happy coincidence from an effort to relate her teaching of mathematics to the teaching of English of a colleague, Mrs. Louise Anderson MacDonald, who taught for many years in Pennsylvania State Teachers College, Indiana, Pennsylvania. Professor Walker relates the anecdote as follows:

It was one of those incidents which could not have been planned and which came to light only through an unusual coincidence. I was teaching mathematics in the Oread Training School of Kansas University in Lawrence and Mrs. Louise Anderson MacDonald held a similar position in the Department of English. We arranged for her to visit some of the mathematics classes in order that she might see what were the difficulties encountered by her pupils in oral exposition. I have no idea what particular geometry problem we were studying on this occasion but I am sure that it was one for which the general theorem had not been stated. Let us suppose we had started with the simple problem of knowing that in a triangle ABC, angle A equals angle B, and that the class had been exploring to see what other properties of the triangle could be deduced from this known fact. The conversation probably went something like this:

Teacher: "Can you tell us now what it is you have found out about the triangle?"

Pupil: "Yes, I started out knowing that angle A and angle B were equal and I have found out that whenever that is so, line AC must be equal to line BC."

Teacher: "That statement is entirely correct but is rather limited in its usefulness. If you want to make use of that same information tomorrow or at some later time, you will need to have it stated in a more general form without reference to the particular letters used to name the triangle."

Then after some prodding, either this pupil or the class achieved the general statement of the theorem. There was nothing unusual about this class session and no one concerned would ever have remembered it except that a few weeks later Mrs. MacDonald asked a student in an English class what was the point of *Silas Marner*. The student replied: "Silas was accused of taking some money which he really had not stolen so he ran away. It didn't do him any good, though." Then to her delight, one of the girls who had been in the geometry class Mrs. MacDonald visited, remarked: "That is true enough but it isn't very useful. If you want to make use of that same idea again in another situation, you ought to generalize it. You might say something like this: When one is accused of a crime which he has not committed, it is quite useless to try to escape by running away."

Bibliography

1. BROWNELL, WILLIAM A. *Learning as Reorganization: An Experimental Study in Third Grade Arithmetic.* Durham, N. C.: Duke University Press, 1939. p. x + 87.
2. BROWNELL, WILLIAM A. "Problem Solving." *The Psychology of Learning.* Forty-First Yearbook of the National Society for the Study of Education, Part II. Chicago: The University of Chicago Press, 1942. p. 415–43.
3. GATES, ARTHUR I. "Connectionism: Present Concepts and Interpretations." *The Psychology of Learning.* Forty-First Yearbook of the National Society for the Study of Education, Part II. Chicago: The University of Chicago Press, 1942. p. 141–64.
4. HENDRIX, GERTRUDE. "A New Clue to Transfer of Training." *Elementary School Journal* 48: 197–208; 1947–48.

5. HENDRIX, GERTRUDE. "Prerequisite to Meaning," *The Mathematics Teacher* 43: 334–39; November 1933.
6. JUDD, CHARLES H. *Education as Cultivation of the Higher Mental Processes.* New York: Macmillan Co., 1936. p. vii + 206.
7. JUDD, CHARLES H. "The Relation of Special Training to General Intelligence." *Educational Review* 36: 28–42; 1908.
8. KATONA, GEORGE. *Organizing and Memorizing.* New York: Columbia University Press, 1940. p. xvi + 318.
9. LUCHINS, ABRAHAM S. and LUCHINS, EDITH H. "A Structural Approach to the Teaching of the Concept of Area in Intuitive Geometry." *Journal of Educational Research* 40: 528–33; March 1946.
10. MCCONNEL, T. R. "Recent Trends in Learning Theory." *Arithmetic in General Education.* Sixteenth Yearbook of the National Council of Teachers of Mathematics. New York: Bureau of Publications, Teachers College, Columbia University, 1941. p. 268–89.
11. NATIONAL EDUCATION ASSOCIATION. New York: American Book Company, *Report of the Committee of Ten on Secondary School Studies*, 1894. p. xii + 249.
12. ORATA, PEDRO T. "Recent Research Studies on Transfer of Training with Implications for the Curriculum, Guidance, and Personnel Work." *Journal of Educational Research* 35: 81–101; October 1941.
13. SNODDY, G. S. "An Experimental Analysis of a Case of Trial and Error Learning in the Human Subject." *Psychological Monographs*, 20: 124; 1920.
14. THORNDIKE, E. L. *The Psychology of Arithmetic.* New York: The Macmillan Co., 1922. p. xvi + 314.
15. THORNDIKE, E. L., and WOODWORTH, R. S. "Influence of Improvement in One Mental Function upon the Efficiency of Other Functions." *Psychological Review* 8: 247–61; 1901.
16. WERTHEIMER, MAX. *Productive Thinking.* New York: Harper Brothers, 1945. p. x + 224.
17. WHEELER, RAYMOND, and PERKINS, FRANCIS T. *Principles of Mental Development.* New York: Thomas Y. Crowell Co., 1932. p. xxvi + 529.
18. WOODWORTH, R. S. *Experimental Psychology.* New York: Henry Holt and Co., 1938. p. xi + 889.

8. Problem-Solving in Mathematics

KENNETH B. HENDERSON AND
ROBERT E. PINGRY

THE present chapter on problem-solving in mathematics is written on the assumption that mathematics teachers should understand the basic theory of problem-solving which is derived from research in the subject and also see clearly the implications of this theory for methods and procedure in the classroom. Both are necessary. Theory apart from the implications and consequences is largely sterile. Methods and procedures apart from a conceptual framework become little more than a bag of tricks. Accordingly, the chapter can be considered as divided into two parts. The first part discusses the theory of the process or group of processes of problem-solving. The second part discusses the implications for classroom procedure. It is hoped this chapter will prove helpful to teachers as they try to help their students not only solve particular problems, but also improve generally in their techniques of solving problems.

WHAT IS A PROBLEM?

One concept of problem, which is a very common one, is that of a question proposed for an answer or solution. It is this concept that the teacher has in mind when he says to his mathematics class, "Your assignment for tomorrow is to work problems one to ten on page 164." The question which may be either explicit or implicit in each problem is, "What is the answer?"

The concept of a problem as a question is the one we have in mind when we speak of educational problems like teaching problem-solving, teaching for transfer, maintaining discipline, providing adequate educational guidance. In these examples the question is implied. It might be phrased as "How can I ... ?" or "How can we ... ?"

A second concept of a problem still considers the existence of a question to be necessary, but unlike the first concept, existence of the question is not regarded as sufficient. The additional conditions pertain to the individual who is considering the question. What may be a problem for one individual may not be a problem for another. A problem for a particular individual today may not be a problem for him tomorrow.

How a person solves a problem. It has been mentioned that a necessary condition for a problem to exist *for a particular individual* is the existence of a question. To identify other necessary conditions, it is desirable to analyze the psychological process of problem-solving; that is, how a person goes about solving a problem.

The first identifiable part of the process of solving a problem is the on-going, sustaining activity of the individual. If we were to probe into the causes of this behavior, we would always find a more or less rationalized goal or an unresolved psychosomatic tension present. It is this goal or tension that causes and directs the individual's behavior. A student, for example, decides to do his mathematics assignment. He finds a place to work, takes a sheet of paper, and begins to work the assigned exercises. This student has a goal in mind. He wants to complete his assigned work.

The behavior of an individual may also be caused by a less clearly defined goal. To use the student again as an example, he completes the assignment and has some time on his hands. He listens to the radio a while, finds he has read all his comic books, looks out of the window, calls a friend on the telephone—all in an attempt to relieve his feelings of boredom. He is under the impact of a tension. but he has not clearly defined the goal which will resolve this tension.

It is the difference between the given situation and the desired situation (goal) that evokes, directs, and sustains the individual's behavior. Other factors being equal, the clearer the individual is about his goal, the stronger is his will-to-do or motivation. A vague feeling of uneasiness is not conducive to behavior which will remove this feeling. A consciously held and clearly defined goal, on the other hand, helps the individual select and organize behavior so that there is greater likelihood of the goals being reached. Attainment of the goal by the individual is satisfying. Tensions are released, the individual's ego is enhanced, and he feels better.

The second identifiable part of the process of problem-solving consists of a blocking of the behavior normally employed by the individual in attaining his goal. The blocking has to be of such a nature that well-established habits cannot immediately go into action to circumvent or remove it. Suppose a girl is considering the problem, "A recipe for making four dozen cookies calls for 1 cup sugar, $\frac{1}{2}$ cup sweet milk, $\frac{1}{4}$ teaspoon baking soda, and 2 cups flour. How much of each ingredient shall I use if I want to make two dozen cookies?" If she immediately takes one-half of each amount, there really has been no blocking. But suppose she does not know what one-half of one-half and one-half of one-fourth are. Blocking has now occurred, and she has become aware of a "problem" in the sense of a question to be answered.

The third step in problem-solving is now ushered in. (This is assuming the student continues to act in terms of his original goal. Should he decide to abandon the assignment, then he has changed his goal. There is no blocking and hence no problem.) The student begins to think, and to figure out ways of removing the block and thereby to attain his goal.

This analysis of the process of problem-solving allows us to identify the necessary conditions for the existence of a problem-for-a-particular-individual:

1. The individual has a clearly defined goal of which he is consciously aware and whose attainment he desires.

2. Blocking of the path toward the goal occurs, and the individual's fixed patterns of behavior or habitual responses are not sufficient for removing the block.

3. Deliberation takes place. The individual becomes aware of the problem, defines it more or less clearly, identifies various possible hypotheses (solutions), and tests these for feasibility.

Implications for the meaning of a problem. The second concept of "problem" discussed here holds that when these three necessary conditions are met, a problem exists for the particular individual. It can be seen that this concept differs from the former one. Every question that is proposed for solution is not a problem. As Cronbach (7:34) points out, "... it is not posing the question that makes the problem, but the person's accepting it as something he must try to solve." Furthermore, a question such as "What is the value of x in the equation $ax + by = c$?" is no problem to someone who understands algebra so well he operates only on the basis of habit in solving the equation. If there is no discrimination among alternative courses of action, no problem exists *for the particular individual*. The question, however, might be a problem to a student who understands how to solve numerical equations but has never solved a literal equation. Such a student cannot depend upon a fixed habit of solving equations. He will be successful when as the result of thinking he is able to identify the general principles involved and apply them correctly.

It is not a question of *which* concept is the correct one; i.e., problems existing independent of persons who might face them, or problems as only existing relative to the persons who face them. It is rather a question of which is more useful for a certain purpose. The second concept of a problem appears to be the more useful concept in most educational contexts. It is the one accepted in the present chapter.

Are textbook "problems" problems? Textbook "problems" may be defined as all kinds of pre-formulated "problems" whether they appear in textbooks or are prepared and presented by the teachers in review exercises or tests. To simplify matters it will be assumed that students can understand the "problems." Of course, if the students are of low mentality, or do not know what the mathematical symbols mean, or see no relationships whatsoever, the "problems" have no chance of becoming problems. They remain enigmas *for those particular students*.

Whether textbook "problems" are problems can be determined by examining them in terms of the three necessary conditions stated in the previous section. When this is done, the only conclusion is that it all depends on the student's reaction. If he accepts the "problem" as his own (that is, if his ego becomes involved), then the solution of the "problem" becomes his goal. In this case the textbook "problem" has met the first condition for a problem

It is important to note that it really makes no difference whether the student poses the "problem" (question) for himself or whether the teacher or textbook poses it. The crucial factor is the extent to which the student's ego becomes involved in the problem. Some educators seem to believe that a "problem" is a problem only if the student formulates it for himself with little or no help from the teacher. Such a position is hard to defend even on theoretical grounds. It remains to be proven that a mathematics teacher who is highly enthusiastic, has an exciting personality, and is a student of psychology, cannot involve students in more problems than can a teacher who waits for the spirit to move the students. The main reason that so many textbook "problems" never become problems for the students is that the teacher does not make much effort to challenge the student. As Bakst (1:9) says, "A properly formulated challenge will, by and large, rarely go unanswered." There are possibilities in pre-formulated "problems." Whether they are realized depends to a great extent upon the teacher.

But assuming that a student makes the solution of the "problem" his goal, there is still the possibility that the "problem" may be easy for him. The solution may be merely a matter of grinding out the answer. Again it depends on the student. There are probably students in every mathematics class for whom only the first few "problems" in an assignment really are problems. These students are sufficiently gifted that they rapidly discover principles which serve to remove the problem-nature of the last "problems" in the assignment. There are probably also students for whom every "problem" in the assignment is a problem; some problems are so effective in setting up blockings that the student is unable to eliminate them and solve the problems. In short, what is one student's problem is another student's exercise, and a third student's frustration.

In the foregoing discussion, no distinction was made between so-called verbal "problems" and ordinary exercises. This was because such a distinction is of no particular value in light of the postulated conditions for the existence of a problem. Verbal "problems" and exercises differ in nature, kind of abilities demanded, and difficulty. But one may be no more or less a problem for a student than the other. It all depends on the student's orientation

THE IMPORTANCE OF PROBLEM-SOLVING

If life were of such a constant nature that there were only a few chores to do and they were done over and over in exactly the same way, the case for knowing how to solve problems would not be so compelling. All one would have to do would be to learn how to do the few jobs at the outset. From then on he could rely on memory and habit. Fortunately—or unfortunately depending upon one's point of view—life is not simple and unchanging. Rather it is changing so rapidly that about all we can predict is that things will be different in the future. In such a world the ability to adjust and to solve one's problems is of paramount importance.

The case for teaching students how to formulate and solve problems involving quantitative thinking is abundantly clear. Come graduation and/or employment, they will have to be able to solve the problems posed for them in their advanced education or in the job they hold. To be able to do this is the pay-off of their education. Because most, if not all, mathematics teachers hold this position, they expend considerable effort in teaching their students how to solve problems in the field of mathematics. There are few, if any, kinds of instruction that potentially have more value.

Should mathematics courses contain more problems? From what we know about learning, there is only one way students can learn to solve problems—by solving problems and studying the process. This means that students will have to be faced with *problems*. Perhaps one of the reasons some teachers have done so badly in teaching problem-solving is that they have confronted their students with or helped them formulate few real problems. This may not be the only reason, however. Unless students study the process of solving problems as an end in itself there is scant likelihood that they will learn the generalizations which will enable them to transfer their ability to solve problems to new problems as they arise. Woodruff emphasizes the importance of studying the problem-solving process directly when he states (28: 301):

Furthermore, in face of what is known about the relative absence of transfer of training in most school subjects, it is the height of folly to expect students to develop problem-solving skill as an incidental learning unless considerable time and attention is devoted directly to it, in which case it ceases to be incidental. It is far more likely that something about civic affairs will be learned in a unit on problem-solving, than that problem-solving skill will be developed in the typical unit on civic affairs.

Before the process can be studied effectively, the mathematics course must contain many problems that fulfill the necessary conditions already identified. It is not a question of either using pre-formulated "problems" or using "life problems." Both have their place. However, mathematics teachers have been inclined to ignore the latter. These are the problems that emerge from social situations, industrial activities, or the personal life of the students and whose solution requires a substantial amount of quantitative thinking. Hartung (12) cites some of the salient characteristics of these problems: (a) They have no definite question, but rather the question(s) have to be formulated at the outset; (b) The necessary data are not given, but rather have to be collected and evaluated; (c) Analysis and interpretation are much more complicated; and (d) A definite answer often is not possible; verification is possible only by actual try-out.

It is much more difficult to find problems of this kind. They probably will not be too satisfactory if in a textbook, for they depend to a large extent upon unique factors in the school in which they are studied. Yet, if such problems do not find a place in the mathematics courses which lend themselves to their inclusion, it is not likely that students will become competent in solving them. The evidence on transfer of training does not hold out much hope.

The function of verbal "problems" and exercises. If the teacher selects verbal problems carefully so as to be at the student's level, and if he can get the students to identify themselves with these problems, then the verbal "problems" become real problems. They are probably as useful for teaching problem-solving (though not problem discovery, definition, and formulation) as if they had not been pre-formulated. Once the student's attention is directed to the process they employed in solving the problems and they understand it, the verbal problems provide the practice material on which the students can apply the principles of problem-solving which they have learned. The function of these problems as practice material is the same as that of any practice material. It has been clearly stated by Brownell and Hendrickson (3: 102): "Provision of an abundance of repetitive practice assures learners opportunity to discover their own learning aids (if they want them) and to develop confidence in their ability to react quickly and accurately upon demand."

With the exception of the syntactical form, the chief difference between exercises and verbal "problems" lies in their intended use. Exercises, such as those dealing with the fundamental operations, exponents, radicals, the binomial theorem, and derivatives, are for the purpose of teaching certain mathematical concepts and generalizations. Verbal problems are for the purpose of teaching the generalizations relative to the process or method of problem-solving. These have no necessary relation to a particular kind of mathematics problem; the problem-solving process is essentially the same for all problems. The study of the problem-solving process is the real justification, rather than the utility of the particular problem, for selecting such problems as those based on time-rate-distance, work, mixtures, coins, and business deals.

Even though the primary function of exercises is to give meaning to and provide practice in applying mathematical generalizations and concepts, there is no reason why they cannot be used for the same purpose as verbal "problems"; namely, to provide practice in applying the generalizations which the students have learned about problem-solving. To use exercises for this purpose would require only a shift in attention. Moreover, using exercises for this purpose would result in broadening the meaning (referents) of the generalizations dealing with the method of solving problems. Examples showing how exercises may be used will appear later in the chapter.

In summary, both the exercises and verbal "problems" which appear in practically every mathematics textbook and are used by practically every mathematics teacher have the same function relative to problem-solving. The fact that the verbal "problems" are associated with problem-solving rather than exercises may be due largely to the fact that few teachers realize the different uses to which exercises can be put. It remains to be proven by research whether exercises or verbal problems are the more effective in teaching problem-solving.

ANALYSES OF PROBLEM-SOLVING

As Dewey defines reflective thinking, there is little difference between this and problem-solving. Hence, his analysis of reflective thinking can be taken as an analysis of the act of solving a problem. According to Dewey there are five steps (8: 107–116).

1. Some inhibition of direct action resulting in conscious awareness of a "forked-road situation."
2. An intellectualization of the felt difficulty leading to the definition of the problem.
3. "The identification of various hypotheses ... to initiate and guide observation and other operations in collection of factual material."
4. Elaboration of each of the hypotheses by reasoning and the testing of the hypotheses.
5. Acting on the basis of the particular hypothesis selected in step four, thereby providing the ultimate test.

This idealized analysis describes how a person ought to think were he an automaton governed only by logic. It does not describe the sequence of a real person's thinking. Dewey, himself, states that people's thinking does not ordinarily follow this particular sequence. Studies of how some of our most creative thinkers think show that, if there is a logical pattern, we are not able to discern what it is. In spite of all this, Dewey's analysis is useful in pinpointing stages in the deliberative process.

Johnson (14) gives a slightly different analysis. He identifies three processes or groups of processes which, he says, regularly occur during problem-solving: (a) "Orientation to the problem"; (b) "Producing relevant material, an elaborative function"; and (c) "Judging, a critical function." (14: 202)

Johnson's analysis is oriented more to the psychological processes associated with problem-solving. It is easier to subsume the psychological concepts which have resulted from experimental research on problem-solving under the first two of the three groups of processes Johnson identifies than it is under one or more of Dewey's steps. But when problem-solving in mathematics is considered, Dewey's fourth and fifth steps—namely, elaboration of each of the hypotheses by reasoning and the testing of the hypotheses, and acting on the basis of the particular hypothesis selected—appear to be more fruitful in a conceptual framework than Johnson's third process: judging. This is largely because judging is less a problem (though it still is necessary) in an objective science like mathematics than it is in fields like sociology, political science, or religion, and in the large realm which we call the day-to-day business of living. Since the present chapter tends to be limited to problem-solving in mathematics, Johnson's third process has been replaced by one labeled "testing hypotheses."

The general conceptual framework which has been accepted for consideration in this chapter, the pre-solution period of problem-solving is:

1. Orientation to the problem
2. Producing relevant thought material
3. Testing hypotheses.

The subsequent discussion is organized in terms of these processes or groups of processes. These processes are not clearly delineated one from the other. There is considerable over-lapping and interaction as Johnson, himself, indicates. "Problem-solving begins with the initial orientation and ends with the closing judgment, but between these bounds almost anything can happen, in any sequence" (14: 203).

Orienting to the problem. Johnson defines orientation in this context as "the process by which the organism grasps the material of thought and keeps it available for deliberation" (14: 204). This process or group of processes is, therefore, an encompassing one and includes the other two: producing the material of thought and testing hypotheses.

An individual's orientation to a problem depends in part on his physical and mental condition. A student who has a violent headache, one who is grossly undernourished, or one who has been up until 2 A.M. can hardly be expected to be effective in orienting himself to classroom problems. A student whose dog has just been killed or one who is keyed up for the football game after school also will probably be ineffective.

An individual's orientation to a problem depends also on what the problem means to him; that is, whether he understands the words and symbols in the problem and their relationship to each other. This orientation depends upon how the problem is related to consciously-held motives, and how it affects such ego-needs as the need for success, approval, belonging, and security. He is aware of the first two of these; i.e., the language of the problem and the problem's relation to his motives, but is largely unaware of the third.

It is postulated that meaning is a continuous variable whose range is the open interval zero to infinity. This means that meaning is not an "either-you-have-it-or-you-do-not" proposition, but rather a matter of degree. Furthermore, it means that every problem has some meaning for an individual.

To illustrate the different meanings a "problem" may have, let us imagine we are watching a group of students attack a problem in their mathematics assignment. To Bob the problem means a short delay until he can get back to his comic book. Hence, his goal is to get *an* answer—not necessarily the right one—which he can use to convince himself that he has completed the assignment. Then he can get to that comic book. To Harriet the problem means a chance to get the attention of the boys in the class and show up the other girls. Hence, her goal is to solve the problem in such a "neat" way as to receive the admiration of the boys or the teacher. To Frank the problem means another threat to his self-esteem; all mathematics problems are. Frank just cannot take failure; he has set a level of aspiration that is so inflexible that it cannot be tempered by his varying aptitudes. Hence, Frank's only recourse is not to attempt the problem. Otherwise, he might fail. To Tom the problem means almost nothing. Words like per cent, discount, and net price have no referents for him. Unlike Frank, however, Tom does not let a little thing like a mathematics problem bother him. Both he and his teacher are used to his saying "no" when asked whether he was able to do the problems in the assignment. To Shirley the problem means "a kind of problem I had better learn to solve if I expect to hold a job in merchandising." She understands the words in the problem and starts to find the answer and also tries to memorize how problems of that kind can be solved.

Here we have five students each facing a "problem." Because of their different motivations, the "problem" means something different to each one. Hence each one's material of thought and plans of action will be different. Many teachers are unaware of all this. Perhaps Cronbach's admonition (7: 35) ". . . the motives which the pupil brings to the mathematics assignment are rarely related to the problem itself," if remembered will be of considerable help to teachers as they work with the students.

Clarification of the problem by the individual is a process which affects his orientation. This is not a step or process that is clearly distinguished from a prior or subsequent step. In a sense it is a continuous process. Each of the five students described above clarified the problem to some extent as he faced it. Each knew, for example, that it involved mathematics, that he probably would have to perform one or more mathematical operations, and that the answer would probably be expressed in numbers.

Further clarification of a problem depends on the student's understanding of the words and symbols used in the statement of the problem. To the extent that he understands these, he is better able to decide what must be done to obtain the answer. In addition he must be able to identify what is given and what is required. If the problem is not pre-formulated, there is still the matter of the perceptions and concepts which form the "given" and some idea of the characteristics of an adequate solution. Duncker (9: 35) points out the relationship of the given and the solution: "A solution always arises out of the demands made by what is required on what is given." With these two points in mind, the student is better able to organize his efforts and secure a solution.

Duncker (9) has introduced a concept that is useful in understanding the psychological process of solving a problem. It is the concept of a "search model." The search model evolves as the individual clarifies the problem. It bridges the gap between what is given and what is required, and serves for a period of time to direct or channel the individual's deliberation. To take a simple example, suppose you meet a friend who tells you something you may forget if you do not make a note of it. You have no pen or pencil. In this case you have something you want to remember—you need a device to help you remember it. Your search model is a mental construct—"something-to-write-with." It is an abstraction obtained from the total situation which contains many irrelevant and confusing elements. But it provides the stimulus that starts you thinking and acting, directs what you look for, and tells you when you are through. It also determines the "region of search," in this case your perceptual field.

To make a more involved problem, suppose a student is considering the problem: Each of two men travels 500 miles after leaving a certain town at the same time. The first man travels 10 miles an hour faster than the second and so arrives 2.5 hours earlier at his destination. How fast did each man travel? Assuming that the student understands the problem and accepts it, his search model—if he verbalizes it—may be "an equation that relates the variables in the problem."

His region of search consists of the mathematical concepts and generalizations he has learned. These are appraised and selected in terms of their usefulness in attaining the search model. If they provide any hunches or suggestions for plans of procedure, the plans are tested to see whether they provide a solution to the problem.

All these concepts, generalizations, and hypotheses, constitute the thought material for the deliberative process. They will be considered at greater length in the following section. The important idea at this point is that of a search model. Part of the teacher's work consists in helping students conceptualize functional search models as they clarify problems. Faulty search models undoubtedly are one of the chief causes of the errors students make in solving problems.

Producing relevant thought material. The second aspect of problem-solving being considered is producing relevant thought material. Such material consists of perceptions obtained directly and immediately from the existing situation, concepts, and generalizations. The latter may either be recalled or obtained from someone else during the process of solving the problem. Whether perceptions or concepts and generalizations predominate depends on the nature of the problem. Spatial perception probably plays a larger role in the problems associated with intuitive and demonstrative geometry, numerical trigonometry, and analytic geometry than it does in the problems associated with arithmetic, algebra, calculus and higher analysis, at least as these subjects are ordinarily taught.

Mathematicians are well aware of the role played by concepts and generalizations in the deliberative process in problem-solving. It is these abstractions which makes it possible to restructure or reorganize past experience and bring it to bear on the problem at hand. There is no substitute for an understanding of relationships manifested by the possession of concepts and generalizations. Few teachers would disagree with this.

The production and retention of thought material is dependent on the individual's "span of apprehension," sometimes called "immediate memory span." The individual must be able to remember what the "given" is and what he is expected to find as he continues to work on the problem. He must also be able to select from his past learning whatever is relevant to the problem.

The span of apprehension has two dimensions: extent and duration. These vary from individual to individual. Some people can readily recall facts, principles, concepts, definitions, theorems, etc. The extent and duration of their span of apprehension is considerable. Others, for one reason or another, have poor memories and have difficulty in making use of their previous learning. Some even have difficulty in grasping all the relationships in the problem or in remembering the unsuccessful things they have done in trying to solve the problem.

The concise symbolism in mathematics is of great aid in increasing one's span of apprehension. By symbolizing the concepts and relationships of a problem, these can be dealt with as entities and kept immediately available for a study of their relations with each other. Hypotheses can be readily tested, and new insights obtained as the symbols are manipulated.

A particularly useful concept in considering the production of thought material is that of "set." A person is said to have a "set" toward a problem when, because of past experience, he is pre-disposed to a particular search model, hypothesis, or plan of action and steadfastly maintains this predisposition. Suppose a class is studying how to find what per cent one number is of another, and is given only exercises which yield a per cent less than 100. Some students may decide that the first step in finding the answer is always to divide the smaller number by the larger. These students probably will develop a set toward such problems and mechanically apply this procedure to all problems involving finding a per cent.

Though a set is not necessarily an obstacle to problem-solving, the term as ordinarily used in accounts of research or explanations of the theory of problem-solving has a negative tone. It carried connotations of rigidity, inflexibility, blocking of the reorganization of experience and the restructuring of a configuration, and a mechanized state of mind.

Some of the most convincing research on "sets" has been done by Luchins. In one experiment he gave his subjects a group of seven problems involving finding a stipulated volume of fluid by manipulating three containers. All seven problems were solved by the formula $a - b - 2c$. For example, if the subjects had a ten-quart measure full of liquid and two empty measures, one holding five quarts and the other two quarts, and were required to obtain exactly one quart, they would fill the five-quart measure and the two-quart measure twice. This would leave one quart in the ten-quart measure. The formula was not known by the subjects.

The problems were solved at first by the process of approximation and correction, to use Melton's phrasing. The first problems required more time, but the last problems were solved forthrightly indicating learning. Then five problems similar in nature at first glance were given the subjects. But all these were simpler, and could be solved by either $a - c$ or $a + c$. The "set" the subjects had attained toward the problem from their success with the initial seven problems pre-disposed all of them to go through the longer $a - b - 2c$ process. Moreover, they persisted in this even though it failed to provide a solution. A second group of subjects not given the first group of problems designed to induce the "set" did not manifest a "set" on the test group of five problems.

Luchins used the name "Einstellung" for this kind of "set." It is defined as "the set which immediately predisposes an organism to one type of motor or conscious act (15: 3)." Under the impact of an Einstellung, a person does not look at a problem on its own merits, but tries mechanically to employ a previously learned method. Further conclusions of Luchins were: (a) the greater the success on the practice set, the greater was the force of the Einstellung; (b) all age groups were affected; and (c) education, as measured by the number of years of schooling and IQ, had no significant effect in reducing the Einstellung. Luchins found from analogous experiments using a geometry theorem that the conclusions held. Reid (24) came to the same conclusions using a different paradigm.

In another connection Luchins and a collaborator said (16: 286):

They [the students] were accustomed to the use of isolated drill in arithmetic, wherein in order to "learn" a method or formula they practiced it in a series of similar problems—a situation quite similar to our experimental setup. They were accustomed to being taught a method and then practicing it; to have to discover procedures was not only quite foreign to them in arithmetic but also in most school subjects. It seems to us that the methods of teaching to which they had been subjected tended to develop, not adaptive responses, but fixations, so that a child might know methods and formulas and yet not know where to apply them or how to determine what method best suited a particular problem. Our schools may be concentrating so much on having the child master the habits, that the habits are mastering the child.

In the present context of a discussion of producing thought material, the relevance of an Einstellung is that such a set materially affects the search model, narrows the range of search, inhibits the perception of certain relationships, and blocks certain hypotheses. The result of this is that not all relevant thought material is made available.

Hadamard (10) suggests utilizing an "incubation period" as one way of breaking a set toward a problem. One lays the problem aside and forgets it. The ensuing incubation period is a period of no conscious work of the mind. However, the subconscious mind (Hadamard postulates the existence of such an aspect of mind) is at work and often insight or illumination occurs. Many people have had the experience of a solution to a problem popping into the mind without any apparent reason. Some have had insights occur during sleep. Poincare (22: 387–388) tells of an interesting experience with problem solving:

Just at this time I left Caen, where I was living, to go on a geological excursion under the auspices of the school of mines. The changes of travel made me forget my mathematical work. Having reached Coutances, we entered an omnibus to go some place or other. At the moment when I put my foot on the step the idea came to me, without anything in my former thoughts seeming to have paved the way for it, that the transformations I have used to define the Fuchsian functions were identical with those of the non-Euclidean geometry. I did not verify the idea; I should not have had time, as upon taking my seat in the omnibus, I went on with a conversation already commenced, but I felt a perfect certainty. On my return to Caen, for conscience' sake, I verified the result at my leisure.

Kekule's idea of the carbon ring for describing the molecular structure of benzene came suddenly when the six carbon atoms appeared to be dancing and then joined hands and formed a ring.

Hadamard held a forgetting hypothesis to account for flashes of insight during the incubation period. The mind gets rid of false leads and hampering assumptions, and approaches the problem less structured by the set induced by these. Poincaré (22) held this same hypothesis. Helmholtz however, believed the explanation lay in the absence of nervous fatigue. (See 29: 818.) With a refreshed mind, ideas can more easily be brought into relation with each other. Whatever hypothesis is accepted, an incubation period is useful in helping break an unproductive set.

Forming and testing hypotheses. A hypothesis, to use Dewey's phrase (8), is an idea that suddenly pops into mind suggesting how a problem may be solved. A student who is contemplating how to factor $4n^2 - a^2 - 2ab - b^2$ suddenly thinks that the expression may be the difference of two squares even though it is not exactly like $a^2 - b^2$ which he is used to. This idea or hypothesis represents an insight—a perception of a relationship or set of relationships not previously seen. Though occurring spontaneously, it has its source in the student's past experience. What it amounts to is a reorganization or restructuring of experience.

Whether a hypothesis is effective or ineffective in solving a problem can be determined either by actually testing it out or by elaborating its consequences and implications in an attempt to uncover any error in fact or inconsistency with tested knowledge. In the student's case, his hypothesis was an opportune one. By acting on it he could solve the problem. Not all hypotheses "work." Those that do not "work" have to be abandoned and the search continued for one that leads the way out of the difficulty.

The broader and deeper is the individual's knowledge and understanding of the subject at hand, the more readily will he form hypotheses. This is assuming that he has no emotional block on the problem or has not developed an Einstellung of one kind or another. One of the obvious differences between a student who solves problems readily and one who does not is that the better student has many ideas concerning what might work. The poor student is rather helpless after his one or two hypotheses prove ineffective. The better student is able to conceptualize readily because he has a greater background of understanding.

It is significant that knowledge *and* understanding were identified as determinants of hypotheses. Knowledge, in the sense of remembering a large number of facts, does not necessarily guarantee fruitful hypotheses. This is amply borne out by experience.

Thorndike (26: 201f) believes there are two reasons for the fact that acquiring knowledge and being able to apply it when appropriate are different abilities. An individual may have an aptitude for the former and not for the latter. Also the manner in which the knowledge is acquired has a bearing on how readily it can be used. Knowledge acquired largely by memorization, by the "pouring-in" process, and without many relationships being established with the individual's existing knowledge has low transferability. Knowledge which becomes understanding by virtue of teaching which fosters learning by discovery, which deliberately establishes relationships, and which aims at broad concepts and generalizations has higher transferability. The ability to transfer one's knowledge and understanding—to find a meaning in a situation—facilitates the formation of hypotheses.

We need to know much more than we do about the psychological process of forming hypotheses. There is room for much experimentation. We do know, however, from experimental studies that giving the students a pattern of "steps" to follow does not significantly help them solve real problems. In fact, Hanna (11) found that when students were left to form their own hypotheses, their success in solving arithmetic problems was not significantly less than that of students who were taught more or less formalized procedures. Until more evidence is forthcoming, teachers will probably not go astray by emphasizing relationships, helping students organize their knowledge, and providing an abundance of opportunities whereby students apply what they have learned. Such teaching, from what we know now, facilitates the formation of hypotheses.

The search model determines to a considerable extent the hypotheses that are formed. As had been explained above, the search model, whether it is a mental picture or a verbalized or unverbalized statement, describes the key which will unlock the problem. In one sense it is a hypothesis, for it represents an inference or guess concerning how to move from the given to the required solution. But it is very general and needs analysis and elaboration. This process of analysis and elaboration gives rise to the hypotheses concerning how the search model can be attained.

Take the case of a pilot of a private airplane who finds that the distance from St. Louis to Chicago on a certain map is 16.1 inches. He knows the representative fraction of the map is $\frac{1}{1,000,000}$, but is not sure how to find the distance in miles between the two cities. His search model is: "a way of relating the distance on the map and the representative fraction." This gives rise to the hypothesis that he can set up some kind of proportion, assuming he knows the meaning of representative fraction.

Once a hypothesis occurs the next step is testing the hypothesis. In mathematics this is done by deducing its implications. The quantitative magnitudes are symbolized, and the symbols manipulated by following the rules of mathematics and logic. If this procedure does not result in the solution of the problem, the hypothesis is abandoned and another sought. A student has a hunch that he can solve the equation $3^x = 10$ for x by taking the xth root of both sides. He applies the appropriate principles and finds that his hypothesis is fruitless.

Sometimes a different form of deduction is used. A student is considering a common denominator of $\frac{3}{4}$ and $\frac{2}{3}$. The number 12 occurs to him as a possible answer. He reasons that *if* 12 is a common denominator, *then* it will be divisible by both 4 and 3. He verifies this. Since a common denominator will be divisible by both denominators, 12 must be a common denominator.

Checking the answer obtained in solving an equation is another example of this "if-then" reasoning. The student reasons, or should reason, that *if* his answer is correct, *then* it should check when substituted for the unknown(s) in the original equation. He performs this check and verifies or does not verify his prediction.

This form of testing is sometimes called prediction and verification. The individual accepts the hypothesis conditionally. He uses "if-then" thinking to predict what will follow. He then ascertains the existence or non-existence of the predictions. If they are verified, the hypothesis is accepted as being the answer to the problem. The necessary assumption should be noted; namely, that the predictions can occur *only* if the hypothesis is true. Should this assumption be false, the method of testing an hypothesis by prediction is not valid.

HOW MATHEMATICS TEACHERS CAN HELP STUDENTS IMPROVE IN PROBLEM-SOLVING

How can mathematics teachers help their students improve in problem-solving ability?

In the previous sections of this chapter it has been shown that problem-solving is a very complex process. In fact, psychologists find it difficult to distinguish between problem-solving and learning generally. From this point of view there is a sense in which this entire Yearbook, rather than just this chapter concerns problem-solving. Motivation, attitudes, transfer of training, drill, concept formation, language, and logic, are all aspects of problem-solving. A teacher who is seeking to improve problem-solving ability must necessarily give proper emphasis to each of these aspects. A program of instruction, however, that involves these necessary phases of learning is not sufficient. It is important that specific experiences designed to foster problem-solving abilities be provided in the program of instruction.

Some people are endowed with native abilities which enable them to solve problems better than others. However, it is possible to help each individual gain certain abilities and attitudes that will help him be a better problem solver than he would be without these. In the next section of this chapter, techniques and experiences mathematics teachers can use to foster problem-solving ability will be discussed. These techniques and experiences will be discussed under the same heads previously used; namely, orienting to the problem, producing relevant thought material, and forming and testing hypotheses.

Helping Students Become Oriented to Problems

A teacher can help students become oriented to problems by teaching them the meaning of a problem. Every mathematics teacher has had students who did not want to concentrate and reflect on problems. These students will read through a verbal problem in the textbook once and will then, without reflecting, raise their hand and state, "I don't understand the problem," or "I can't solve this problem." Teachers frequently attribute this reaction to laziness when the difficulty may be that the student does not understand what a problem is. May not all the student's past experiences with drill exercises have led him to generalize that problems should be worked without deliberation? He may think one should be able to read a problem and immediately know the answer.

The teacher needs to help this student learn the meaning of a problem. The student needs to learn that a problem is a situation for which one does not have a ready solution. He needs to learn that he is supposed to have difficulty with a problem, he is supposed to have to deliberate. The problem, if he accepts it as a problem, should make him work; if it does not, then it is not a problem for him.

For the student who does not understand the meaning of a problem the teacher may say: "John, you are not supposed to *know* how to work that problem. You are supposed to *figure out* a way to work it. If you have trouble, that's what you should have.

Read the problem several times and draw some diagrams if necessary. Think for several minutes on it before you ask for help."

A mathematics teacher is a psychology teacher. In a sense a mathematics teacher is a psychology teacher teaching the psychology of problem-solving. If students are to improve in problem-solving ability, then the teacher must give the students some guidance in problem-solving processes.

It is important that the students not only know what a problem is, and what some of the aspects of the problem-solving processes are, but why it is desirable to solve problems in school. The teacher's task is two-fold concerning problem-solving. One aspect is that of helping the students with the problems at hand. The second aspect is that of helping the students understand the problem-solving processes per se.

Of course, before the teacher can teach problem-solving the teacher must understand problem-solving. Mathematics teachers need to be students of problem-solving processes as well as students of mathematics. There is considerable evidence that many mathematics teachers do not understand what problem-solving is; or if they know, they do not have it as an objective of instruction. One example of this is the manner in which many teachers teach the verbal problems of the algebra course. Many of the problems are catalogued into types such as "mixture problems," "coin problems," "age problems," and others. The teacher demonstrates to the student how to solve the type, and a list of problems of the type is then given to the students. The students do not experience problem-solving. Rather, they experience practice of applying a memorized technique.

In our culture some situations are faced so frequently that it is desirable to memorize techniques for handling the situation. But are "coin problems," "age problems," and "mixture problems," this important? Problems concerning coins, age, and mixtures can be used to an advantage for improving problem-solving ability. When these problems are taught as memorized types and solutions, however, the opportunity to improve problem-solving ability is lost.

Orientation to problems depends upon a well-organized body of knowledge pertaining to the problem. One aspect of becoming oriented to a problem is to understand the place of the problem in the total organization of the subject. Hence the teacher should take time to provide experiences that will help the student realize the large aspects of the subject as well as the small. Sometimes teachers spend so much time on the details that the student loses sight of the over-all organization. The student cannot see the forest for the trees.

A lesson to help the student understand the historic development of numbers would be very appropriate before starting a unit on complex numbers. This lesson or lessons could deal with the general development and extension of number systems, with attention to their characteristics, and rules of operation. The student would then understand complex numbers in a broader and more meaningful setting.

Some geometry textbooks and teachers help the student outline and organize. For example, the students may be asked to list all the ways they know for proving two angles equal. They may be asked to make a schematic diagram showing the relationships between the areas of polygons, or they may be asked to arrange the members of the set: parallelogram, square, quadrilateral, rectangle, in increasing order of generality.

The review lesson is also a device that enables the teacher to focus the students' attention on the over-all organization of the subject. Review lessons should be planned to accomplish much more than drill on memorizations. The review lesson should help the student organize his past learning into logical and interrelated outlines.

The mathematics teacher is a reading teacher. Verbal problems in textbooks have the additional difficulty over problems arising out of the everyday experiences of the students in that the students must first read a description of the situation before they can be oriented to the problem. Frequently teachers state that the students can't work verbal problems because they can't read. This is a vague statement because the act of reading is very complex, involving many skills and understandings. What does a teacher mean when he says the student can't read? Does this mean the student can't pronounce well, does not read smoothly, does not understand the concepts, or what does it mean? Stating that a student can't read is making use of a catch-all word that may mean any number of things.

Reading verbal problems in mathematics texts does require a different reading technique than reading descriptive material or fiction. The verbal problems are written in a brief, highly compact style, using many technical words. The technical words have to be meaningful to the student before he can understand the problem.

Consider this problem:

A business man was forced into bankruptcy with assets of $15,800 and liabilities of $27,600. What per cent of his liabilities did he pay? One creditor was owed $400. How much did this creditor receive?

What does it mean to state that a student can't read this problem? It may mean that he does not have an understanding of bankruptcy procedure. If this is so, then one way the teacher can help the student read the problem is to help him understand bankruptcy procedure. The student's distraction from being oriented to the problem may also depend on the meaning of one word; for example, the word "liability" or the word "creditor" may cause trouble. The mathematics teacher can help the student read the problem by teaching the student the meanings of these words.

Teachers can also help the students read verbal problems by suggestions such as these: "Mary, you can't read a verbal problem like you read a story. Read is slowly, reread it, read it a phrase at a time, and draw a diagram if necessary to help you remember the important items. Many times one word is extremely important. Do you know the meaning of each word?"

Individual differences exist relative to problem-solving. It was pointed out in the previous section that a problem-for-a-particular individual depends on the individual. For some a situation may be a problem, for others it may be too difficult. A teacher who is trying to help students become oriented to problems must recognize these differences and try to provide each student with problem-solving experience at his level. Some students may sit in a mathematics class all term without facing a challenging problem situation. The assigned textbook problems are routine exercises for them. Another chapter of this Yearbook deals with techniques for taking care of individual differences. It should be noted, however, that the recognition of and provision for individual differences is important in teaching problem-solving.

The teacher can help the students become oriented to problems by encouraging them to verbalize, diagram, dramatize, and construct models. As a student becomes oriented to a problem he essentially has formed a search model. One device a teacher can use to help the student develop this search model is to ask him to tell in his own words what he understands the problem to be. This accomplishes several things. The student, by this experience of telling, is forced to organize his thoughts, and he may clarify certain aspects of the situation for himself. He also may realize his inability to tell about certain aspects of the problem and thus be aware of his weaknesses in understanding. The teacher also has an opportunity to diagnose the student's difficulty. A definite attempt should be made to have the students use their "own words" as use of textbook terminology may not indicate understanding.

The student may make a statement similar to this, "I know this ... and what I want to do is this ..." This verbal statement defines the search model for the student. He has defined the problem, he knows what he wants to accomplish, and is now ready to fill in the gaps.

Diagramming a problem situation is also a very helpful experience to many students. The diagram may be a sketch of the situation or it may be a symbolic diagram. The diagram helps in clarifying relationships between details. The diagram also helps the student keep the many facts and relationships of the problem situation more immediately available than his memory could do. Students should be encouraged to diagram many of the problems they solve. The diagram can also serve as a check on the solution.

Most students in considering this problem should make some diagram or sketch.

A ship is steaming toward Philadelphia at an average rate of 30 miles per hour. It radios that a person is ill aboard and must be picked up. When a plane leaves Philadelphia for the ship, the ship is 270 miles away. The plane averages 180 miles per hour. How long after it leaves Philadelphia does it reach the ship?

PROBLEM-SOLVING IN MATHEMATICS

The diagram may be just a crude sketch to help the student keep the facts in mind as in Figure 1.

Philadelphia   180
•————————————▶   270   30
                                                     ◀———•

FIG. 1

The diagram can be drawn to scale as in Figure 2, and thus give the student a check on his answer.

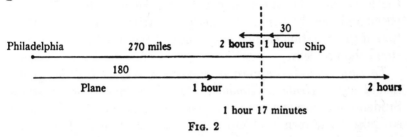

FIG. 2

From this diagram the student can recognize that the plane would meet the ship after about one hour and fifteen minutes.

The student may also want to use a graphic solution, such as shown in Figure 3.

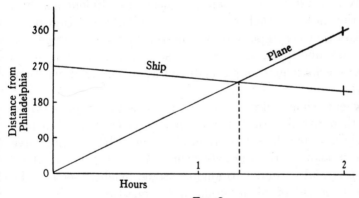

FIG. 3

Some problem situations, especially those problems in three dimensions, become clear to the student if he will construct a model. This model serves essentially the same purpose as the diagram.

Another very useful device to help students become oriented to a problem is to have them dramatize the situation. As the students play the roles of the characters in the problem situation, the problem situation takes on meaning. In an eighth-grade mathematics class some students were interested in the problem concerning a lost, pre-endorsed check, cashed by the finder, but on which payment has been stopped by the maker. The students had considerable difficulty in understanding this problem until they played roles. One student acted the part of the check-maker, one the loser, one the finder, one the store keeper who cashed the check and one the banker. A check was written and a dramatization of the situation carried out. Many of the students now understood the problem.

To help students improve in problem-solving ability the teacher should try to create a climate in the class friendly to questions. Students who are encouraged to ask questions, and feel free to do so, raise their own problems. Many times students fail to ask questions in class even when they have good questions because they are fearful that the teacher will be impatient with them, or that he will not want to take time away from the required work to consider the question. In some classes students and teacher are too ready to laugh at an elementary question a student may ask in all seriousness. A teacher who wishes to build an atmosphere in the class that is sympathetic to questions must encourage students to ask questions, give each question consideration, praise students for asking questions, and discourage laughing at question-askers.

The teacher should also ask many questions that require thinking and then give the students an opportunity to think. Teachers frequently ask questions that require thought, but do not have patience enough to let students think. A student is called on immediately and expected to answer. Some teachers seem to be afraid of a period of silence following a question.

One teacher asks questions this way. "Now think about that. Don't guess. Take some time. I don't want any hands raised for a while. After you have thought and can give evidence that you have, you may answer." Following this admonition there is a period of silence. Sometimes this period of silence may be several minutes long. How different this situation is from the one in which the teacher asks a question, then immediately calls upon John. While John is trying to get his thoughts together and give a reasoned answer, the teacher becomes impatient and turns to James. May not John arrive at the conclusion that in this case one is not supposed to *think*, one is supposed to *know?*

If a thought-provoking atmosphere pervades the class, the teacher may turn many of the students' questions back to them for their own consideration. If the teacher will encourage the students to answer some of their own questions and allow time to work on these questions, then essentially the students pose problems for themselves.

Helping the Students be More Productive of Thought Material

Productivity of solutions of problems depends upon several factors. The principal factors are: the general intelligence level; the background of experience, knowledge, skills, and understandings; the emotions of the student; the motivations; and the field in which the problem is set. So, if a teacher wants to help a student improve in productivity of hypotheses leading to the solution of problems, he must teach so as to give due consideration to each of these factors.

The field in which a problem is set greatly affects the productivity. The time and place and the particular pattern of events leading up to a problem make considerable difference in the productivity of hypotheses. Students in a plane geometry class faced with the problem of determining whether or not two angles are equal may search for two congruent triangles in which these two angles correspond. At another time in the course parallel lines cut by a transversal, or opposite angles of a parallelogram may be uppermost in their minds. The focus of attention upon relationships in a problem is greatly affected by the particular conditions present at the time the problem occurs.

The particular field of events surrounding a problem is somewhat within the control of the teacher. The manner in which the teacher asks a question or the particular time chosen to present a problem may greatly affect the students' search model. One problem may be presented immediately following another problem in such a way that the first problem provides a hint concerning the solution to the second.

One very useful technique that the teacher can use and can encourage the students to use is the heuristic method. The teacher can ask the student questions in such a way that the field will be changed, the student's focus of attention will be changed, or some new element may be brought into the field. This new field may now enable the student to arrive at a solution.

The reader should solve this problem. Construct a circle O having a one-inch radius and construct two diameters AB and CD perpendicular to each other. Now select any point E on the circle and construct EF parallel to CD meeting AB at F and $EG \perp CD$ meeting CD at G. How long is line GF? The reader may find it helpful to solve this problem (if it is a problem for the reader) before reading the following paragraph.

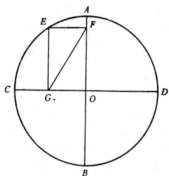

If the reader has experience similar to that of many mathematics teachers to whom the writers have given this problem, the Pythagorean theorem concerning triangle GOF appears in the field and holds promise of a solution. Many students have worked a considerable length of time on this problem without success because they became fixed in their attack on the problem. The focus of attention was the right triangle. The Pythagorean theorem is such a powerful tool that it is reasonable to be diligent in trying to make use of it. However, the able problem solver is one who is able to shift his attention. He realizes the dangers of rigidity and trys to broaden his approach to the problem. The teacher is in a good position to help the student learn something of the psychology of problem-solving and at the same time help the student with this problem by a statement such as this. "When you are having trouble with a problem look at it in a different manner than you have been doing. You have been thinking about triangle GOF, have you tried thinking about the quadrilateral EGOF? Try it. Get in the habit of asking yourself questions about the problem as you work on it."

Polya, in his book, *How to Solve It*, strongly recommends a long list of questions one should ask himself as he tries to solve a problem. These are also good questions for a teacher to ask a student. Some of the questions Polya suggests are (23: inside cover):

1. "Have you seen it before? Or have you seen the same problem in slightly different form?
2. Do you know a related problem? Do you know a theorem that could be useful?
3. Look at the unknown! And try to think of a familiar problem having the same or similar unknown.
4. Here is a problem related to yours and solved before. Could you use it? Could you use its result? Could you use its method? Should you introduce some auxiliary element in order to make its use possible?
5. Could you restate the problem? Could you state it still differently? Go back to definitions.
6. If you cannot solve the proposed problem, try to solve first some related problem. Could you imagine a more accessible related problem? A more general problem? A more special problem? An analogous problem? Could you solve part of the problem? Keep part of the condition; drop the other part; how far is the unknown then determined; how can it vary? Could you derive something useful from the data? Could you think of other data appropriate to determine the unknown? Could you change the unknown or the data, or both if necessary, so that the new unknown and the data are nearer to each other?
7. Did you use all the data? Did you use the whole condition? Have you taken into account all essential notions involved in the problem?"

The result of such a questioning by teacher or student of himself is a shifting of the field in which the problem is set or a shifting of the focus of attention.

The teacher also controls the field of a problem by the manner in which the question is asked or by the particular time the teacher chooses to ask a question. The teacher may present a problem immediately following another which gives the student a useful suggestion of the method of solution of the immediate problem.

Suppose the student is puzzling over the addition of two common fractions whose denominators are unlike. Rather than say, "First find the lowest common denominator," the teacher might ask, "Why don't you add the fractions just as they are?" When the student explains why this cannot be done, the teacher might say, "Then what will you have to do?" The questioning should be provocative of thinking, forcing the student to justify his answers.

Many times a teacher will work out a problem for the class or a particular student. Or he will have some student put the solution on the blackboard. Over twenty years ago Westaway (27) argued against such practices. His argument is still sound. To be sure, the students see how to solve the problem. But as Westaway says (27: 460), "They (the students) are still ignorant as to the way in which the teacher discovered how to solve the problem." A better practice is to ask a question which will direct the students' attention to a key relationship or hypothesis. If they are unable to answer they should be told just enough to enable them to get started. To tell them more destroys the feeling of achievement, and encourages dependence on the teacher.

Sometimes all that is necessary to help a student break a "set" towards a problem is to tell him not to persist in an unsuccessful search model. Maier (18) found this successful in helping his subjects avoid a "set" which was unproductive of a solution. If students learn this as a principle of procedure, the duration of a "set" should be reduced and their productivity of hypothesis increased.

Students should be advised to abandon temporarily the attempt to solve a problem on which they have worked unsuccessfully for a long time, and to return to it later. This provides for an incubation period during which perspective may be restored. Also, if the student has become tired, a rest may be what is needed. When the problem solver returns to the problem the rest or the shift of attention may enable him to be successful.

Many examples can be cited of many famous discoveries that have come in a flash of insight that followed an incubation period or rest period. Many proverbs are present in our culture that encourage recess from working on a problem.

"Take counsel of your pillow."
"If today will not, tomorrow may."
"Sleep on it."

Of course one doesn't solve problems by sleeping, resting, and laying the problems aside. This technique of the incubation period only works after much deliberation and under conditions when the problem solver is strongly motivated to arrive at a solution.

The technique of searching for an analogous problem is a very useful technique for a student to learn. A student faced with a problem in three dimensions may be guided by analogous situations in two dimensions. For example, there are many similarities and likenesses between the geometry of the sphere and the geometry of the circle. Coolidge (6: 226) was impressed with this analogy and stated, "The likeness between circles and spheres extends beyond individual theorems to general methods of proof. Often the procedure which is applicable in one case may be directly transferred to the other."

Polya (23: 38, 101) points out how a problem solver who is concerned with the problem of finding the center of gravity of a homogeneous tetrahedron, and the problem of finding the center of a sphere circumscribed about a tetrahedron could be guided by analogous situations in plane geometry.

Many teachers have long recognized the value of encouraging students to start with the conclusion or the end result and work by analysis to the given. Geometry teachers and textbook authors encourage the students to reason analytically. The student should learn that when he faces a situation for which he has no immediate solution he can profitably direct his thinking by starting with the "to prove" or "conclusion" and, saying to himself, "If I show this I will first have to prove this. This in turn requires that

I know . . . ," until the given data and conclusion are linked logically.

Students should be encouraged to use inductive procedures to help them make discoveries and lead to conjectures concerning the problem at hand. Several specific number relationships may lead the student to discover a general relationship. Many drawings of geometric configurations may lead the student to discover the essential characteristics of a proof. Mathematics is a deductive system, but induction certainly plays a great role in the discovery and creative aspects of the subject.

Continually pointing out and stressing relationships will help students form hypotheses. The fact that a student forms a hypothesis concerning the solution to a problem indicates that he perceives a relationship; e.g., "that follows from this," "this problem is similar to the one I did yesterday," "this is a case of . . ." "the principle that appears to apply is . . ." "If I can find . . . then I can solve the problem." Hence the more a teacher stresses relationships, the better able the students will be to form hypotheses in the subject studied.

Students should have experiences in which they have to identify and define a problem, identify the variables and constants involved, make assumptions which simplify the problem, collect and evaluate relevant data, decide on the characteristics of a satisfactory solution, and finally arrive at a solution which satisfies these characteristics. Problems which call for such abilities are rarely found in textbooks. The author of a mathematics textbook writes a concise, well-structured verbal problem which contains all the data and only those data needed to solve the problem. These kinds of problems serve a purpose, but a steady diet of these will not provide the kind of training needed to solve the real problems in the various occupations and in everyday life. Such problems do not come all packaged and ready for immediate solution.

One way of initiating problems of the kind called for by this principle is to begin with relatively unstructured questions such as, What does your income have to be to afford to get married? Is it cheaper to buy a house than to rent one? How can you save money in buying? and What is the effect on the graph of the quadratic function, $ax^2 + bx + c$, of varying the constants a,

b, and c? Assuming that it is possible to interest students in the problem, the experiences mentioned above follow as normal consequents.

Lund (17) describes a problem initiated by the question, "How does air distance compare with highway distance?" The students used WAC charts which they obtained from the U. S. Coast and Geodetic Survey, a set of U. S. A. highway maps, and ordinary rulers. Before the main problem could be answered, such subproblems as selecting a representative sample, and collecting, presenting, and organizing the data had to be solved. Though the objective seemed to be to teach students some elementary principles of statistics, the possibilities such a real problem offers for simultaneously teaching problem-solving are readily apparent.

Meek and Zechiel (19) describe an insurance company planned and operated by seventh-grade students. This gave rise to many problems most, but not all, being quantitative in nature. Other teachers have had similar success by organizing their classes into companies which engage in retail selling.

Irland and Ensign (13) conducted an experiment on determining automobile stopping distances which might serve as a joint project between a mathematics class and a class in driver training. There are many opportunities for originality and initiative in such a project.

Miller (20) suggests a problem situation involving a pint of water placed on a gas burner. Several functional relations may be identified and studied; e.g., the temperature and time with a high and a low flame, and the time and amount of water remaining.

The project, "What does it cost to own and operate an automobile?" was conducted by Montgomery (21). His class identified the variables involved, collected data, made simplifying assumptions regarding some of the variables, and arrived at a cost per mile for each of the first three years of the car's life. Each student selected a different car.

Many geometry teachers have encouraged students to formulate their own problems by means of flexible models and diagrams (25). By manipulating these models and by studying the flexible diagrams, students may intuitively believe a certain relationship is true. The student then poses the question, "Is this relationship always true?" Careful investigation may show intuition was right or it may show it was wrong. The student, however, has had the experience of formulating and investigating his own problem.

These examples show what can be done in the way of presenting relatively unstructured problems when the teacher is disposed to do so. Without such experiences, students may not get a broad understanding of problem-solving.

Textbook lists of problems should be devised so as to aid in reducing mechanistic, rigid, formula-applying thinking on the part of the students. Luchins (15) made several recommendations to teachers who desire to help their students gain in problem-solving abilities. He suggests that the teacher, after illustrating a method, should not give a set of exercises all alike, but should intersperse problems not solvable by the same procedure. As the result of a later experiment, Luchins and Luchins (16: 293) recommend:

To be effective, problems which aim at conveying the importance of discovering, selecting, evaluating, and discarding facts and hypotheses in solving problems, should be introduced in all school subjects and should not be treated as curiosities which must be heralded with a special introduction, but should be freely intermingled with other more routine problems. If they involve insufficient or additional hypotheses, these should not be patterned as to number or kind. To be sure, as our experiments indicate, the inclusion of such problems may make learning slower and somewhat less efficient than drill procedures, but may also produce less mechanical behavior and more productive thinking. Basically, it revolves on whether our schools wish to develop mechanical efficiency and a formula-applying attitude, conducive to associating a particular method with a particular situation, or whether they aim to develop individuals who have some capability in facing and coping with new and changing problem situations.

Textbooks should contain problems that require the student to be alert and imaginative in solving problems. The student should have the experience of concluding that, under the given conditions, a particular problem has no solution. Rather than problems all solvable by the same technique, there should be some problems that have insufficient data, and some with extraneous data. Some would be absurd questions, and some would be solvable by both long "strong-armed" methods and by short elegant methods.

Recognition should be given to the student who solves a problem in more than one way, and to the student who is able to find a particularly neat solution. Brownell (4: 439) suggested "To be most fruitful, practice in problem-solving should not consist of repeated experiences in solving the same problems with the same techniques, but should consist in the solution of different problems by the same techniques and in the application of different techniques to the same problems. A problem is not necessarily 'solved' because the correct response has been made."

In many mathematics classes the student receives his grade for the course on the basis of *answers* on homework and on tests. May not some students generalize that the process is unimportant just so it gives a correct answer? Students are sometimes satisfied with an incorrect process that results in a correct answer.

Students should become aware that the process of solution is very important. Many problems in mathematics textbooks are solvable by several different methods or devices. A student should develop the habit of trying several solutions. This will help him avoid the mechanized approach of solving a problem by a formula-applying, step-by-step procedure, as well as give the student a good check of his answer. Teachers should give proper recognition and reward to the student who does try several solutions or has searched until he has found an interesting or neat solution. The teacher should ask, "Who has been able to solve this problem another way?" "Which solution do you like best?" "John has a very interesting and brief method of solving this problem. John, will you show us your solution?" Proper reward may also be given in the form of report card grades that include consideration of *processes as well as answers*.

The teacher should develop an emotional climate in the classroom which helps students concentrate on the problems they face. Some teachers employ methods and stimulating devices which divert students' attention away from the problems at hand and toward extrinsic rewards and incentives. The students may have a consciousness of their own inadequacies, or a consciousness of how much they dislike the teacher. Relative to incentives, Luchins and Luchins (16) observed that the children in the experiment were not interested in the problem *per se;* only in the effect the problems would have on their marks, report cards, and whether the principal would know of their performance. These effects were accentuated under speed conditions. The authors conclude (16: 288), "We wonder whether our schools with their stress on grades, their test tensions, and competitive atmospheres, are not conducive to an emotional, highly ego-involved approach toward problem-solving, and consequently, whether they are not detrimental to productivity and flexibility in thinking."

Unwholesome emotional tensions will very probably rest in the classrooms of teachers who use fear as a motivating force; who openly reject students who have difficulty and make mistakes; who continually deflate students' egos by sarcastic, snide, or disparaging remarks; who make unfavorable comparisons of students with their peers; who demand an inflexible standard from all students; or who have no sense of humor and are nervous and irritable. Students in such classrooms will almost certainly find it harder to keep their minds on their work. It may be the dislike for such methods may transfer to the subject and be generalized to include all mathematics.

The principles to be followed in developing an emotional tone in the classroom which does not distract or incense students are rather well established. They can be found in most texts on mental hygiene. Suffice it at this point to mention only a few. A patient and sympathetic attitude will encourage students to try. This is especially true if a teacher does not reject students who cannot measure up to others in solving problems, but accepts them as challenges to his own problem-solving ability.

The old adage, "Nothing succeeds like success," suggests that teachers pace their students carefully; not frustrating and discouraging them by continually setting tasks they cannot accomplish, but starting with things they can do and gradually increasing the difficulty as they gain confidence. Praise, recognition, and encouragement have long been known to relieve emotional tensions. Finally, attention to the interpersonal relations of the members of the class and to the regard they have for each other may serve to diminish feelings of hostility or fear in some of the students. When such feelings are present, it is hard to give one's full attention to solving a difficult problem.

Diagnosis of difficulty and remedial instruction are important. The teacher can help a student become a better problem solver by diagnosis of his problem-solving procedures and provision of remedial instruction. Bloom and Broder (2) conducted an interesting and valuable study on the problem-solving process of college students which included a remedial program, the principal aim of which was to foster general problem-solving ability. Students first were asked to solve problems doing their thinking aloud. A record was kept of the student's remarks as he sought a solution to his problem. Following this experience of solving problems the students next analyzed their own problem-solving methods and compared them to the methods used by their fellow students. The results of this experiment are encouraging to a teacher who desires to help his students improve in problem-solving ability. The students in the experiment made significant gains on problem-solving tests and expressed an increased confidence in problem-solving.

A mathematics teacher can watch the student as he works. Some questions the teacher can keep in mind as he watches the student work are: Does he read the problem carefully, rereading it if necessary? Is he easily distracted? Does he withdraw from the problem and take refuge in daydreaming? Does he write his computations all over the paper in a disorganized manner? Does he check his answers? From all these questions and others the teacher can formulate hypotheses concerning the student's difficulties. These hypotheses can be tested by providing the remedial instruction necessary to remove the difficulties. The technique of having the student work his problems aloud offers some hope in helping the students.

Teachers can help students improve in problem-solving abilities by not requiring step-by-step procedures to be followed. At the arithmetic level there has been considerable research in methods to improve problem-solving ability in arithmetic by various step-by-step procedures. Clark and Vincent (5) devised a plan known as the graphical analysis method. Many textbook authors present a more conventional plan which the student is to follow step-by-step such as: (a) What is given? (b) What is to be found? (c) What operation is to be used? (d) What is the answer? Another method.

called the dependencies method, directs the student to follow the plan of stating "I am to find ———, this depends upon ———." Other standard procedures for solving arithmetic word problems have been advocated. Much research has been conducted to find which method is superior. The findings have been conflicting and at present no step-by-step procedure for solving problems has been shown to be satisfactory for teaching to students. For the students with agile minds, following a fixed sequence is a handicap. For the slower students the sequence may cause them to lose sight of the pattern of relationships involved in the problem. Beyond reading the problem and finding out what is required, there appears to be no fixed sequence of steps.

The teacher should not teach students a specific method which can be used only for a particular problem—or worse, has to be unlearned when other problems are studied. Teaching the students to fill in a box or table in a mechanical way may help them get an answer to a particular problem but it will not help them solve other problems.

Helping Students Improve in Testing Hypotheses

The aspects of problem-solving used in this chapter, (a) orientation, (b) productivity, and (c) testing hypotheses probably do not occur in one, two, three order, but are interwoven into a fabric of thinking in which it is difficult to discern these aspects. However, for the purpose of thinking about problem-solving it is helpful to discuss these aspects. After one has formed a hypothesis, or simultaneously with the formation, the hypothesis must be tested. Our intuition may be of tremendous help in forming the hypothesis or conjecture, but it may also lead us to erroneous conclusions. The principal skills and understandings needed for testing hypotheses are those of inductive and deductive thinking. A large number of data gathering and analyzing skills and understandings usually associated with inductive processes are needed. Understandings and abilities in the use of if-then thinking are needed. Knowledge of logical fallacies is necessary. The student needs to form the habit of suspending judgment until he has systematically studied the many aspects of the problem. He needs to form the generalization that hasty conclusions are frequently wrong.

The technique of testing hypotheses by prediction and verification should be explained to students. This will provide them an understanding of part of the process of solving a problem and will give direction to their efforts.

One of the best ways of illustrating this technique is in the solution of an equation. The "answer" obtained is really a prediction that this answer is a root. When the answer is checked it is verified as a root, or as not a root. Another instance of prediction and verification is the estimation of an answer to a problem before it is worked and the subsequent comparison of the answer obtained with the estimation.

Students should be taught that it is better to form wrong hypotheses provided these are tested and the errors discovered than it is to be without a hypothesis. Sometimes students get the idea that unsuccessful attempts disclose their ignorance. It may be helpful to remind these students of Thomas Edison's reply when he was asked if he were not discouraged after working unsuccessfully so long on a certain invention. Edison is supposed to have replied that he was not at all discouraged as he now knew many things that would not work. If the student remembers all of the things that did not work as he tried them, this in itself should be helpful in arriving at something that will work.

Instead of telling a student that the hypothesis he is contemplating is not valid, let the student find out for himself. For example, suppose a student tries to solve the equation $\sqrt{x} + x = 10$ by squaring each side as the first operation. Instead of telling him he should first subtract x from each side, let him test his own hypothesis. This builds independence. This also keeps the student's attention on the problem where it belongs rather than on the teacher. The student will have the valuable experience of testing a hypothesis that does not work.

As students test hypotheses, some of them ill-conceived, the teacher must be patient and objective lest the student test the hypotheses not by trying them out but rather by watching the teacher's reactions. Many teachers have been discouraged by a student who after reading a verbal problem begins to guess as to the operation to be performed. "Subtract?" the student will ask. Past experience may have taught the student that if he watches the teacher's reactions he can find a clue as to which operation to perform. The teacher can help the student gain better problem-solving experience by giving a noncommittal answer such as "What do you think?" or "Perhaps" or "How do you know?" If the teacher gives such answers independent of whether the answer is yes or no, the student will not get his needed clue from the teacher's reaction. In class discussion also, a student's correct answer may be challenged by other students and much learning takes place if the teacher is not too ready to inform the class that the answer is correct. In general, the student should be forced whenever possible to test his hypothesis by himself.

SUMMARY

Mathematics teachers believe that the ability of a student to solve mathematics problems is dependent upon how deep his understanding of mathematics is. The student's ability to solve problems also depends upon the student's understandings, attitudes, and skills concerning problem-solving processes. This implies that the teacher of mathematics must understand mathematics as well as the psychological processes of problem-solving to be of help to the student. To provide such an understanding of the latter, this chapter attempted to set forth a conceptual framework of problem-solving and point out some of the implications of this for classroom procedure. The hope is that this will afford a teacher fruitful hypotheses concerning his own efforts to teach students the set of understandings, attitudes, and skills conducive to solving problems.

Bibliography

1. BAKST, AARON. *The Problem of Problem Solving in Mathematical Instruction.* Mimeographed pamphlet. New York: New York University Bookstore, 1950.
2. BLOOM, BENJAMIN S., and BRODER, LOIS J. "Problem-Solving Processes of College Students." *Supplementary Educational Monographs*, No. 73. Chicago: University of Chicago Press, 1950.

3. BROWNELL, WILLIAM, and HENDRICKSON, GORDON. "How Children Learn Information, Concepts, and Generalizations." *Learning and Instruction.* Forty-Ninth Yearbook, Part I, National Society for the Study of Education. Chicago: The University of Chicago Press, 1950.
4. BROWNELL, WILLIAM A. "Problem-Solving." *The Psychology of Learning.* Forty-First Yearbook, Part II, National Society for the Study of Education. Chicago: The University of Chicago Press, 1942.
5. CLARK, JOHN R., and VINCENT, E. LEONA. "A Comparative Study of Two Methods of Arithmetic Problem Analysis." *The Mathematics Teacher* 18: 226–33; 1925.
6. COOLIDGE, J. L. *A Treatise on the Circle and the Sphere.* London: Oxford University Press, 1916.
7. CRONBACH, LEE J. "The Meaning of Problems." *Arithmetic, 1948.* Supplementary Educational Monographs, No. 66. Chicago: University of Chicago Press, 1948. p. 32–43.
8. DEWEY, JOHN. *How We Think.* Boston: D. C. Heath and Co., 1933.
9. DUNCKER, KARL. "On Problem-Solving." *Psychological Monographs* 58: 1–111; 1945.
10. HADAMARD, JACQUES. *The Psychology of Invention in the Mathematical Field.* Princeton, N. J.: Princeton University Press, 1945.
11. HANNA, PAUL R. *Arithmetic Problem Solving.* New York: Bureau of Publications, Teachers College, Columbia University, 1929.
12. HARTUNG, MAURICE L. "Advances in Teaching Problem-Solving." *Arithmetic, 1948.* Supplementary Educational Monographs, No. 66. Chicago: University of Chicago Press, 1948. p. 44–53.
13. IRLAND, M. J., and ENSIGN, E. E. "An Experiment in Automobile Stopping Distances." *School Science and Mathematics* 46: 267–71; 1946.
14. JOHNSON, DONALD. "A Modern Account of Problem-Solving." *Psychological Bulletin* 41: 201–29; 1944.
15. LUCHINS, ABRAHAM S. "Mechanization in Problem Solving, the Effect of Einstellung." *Psychological Monographs* 54: 1–95; 1942.
16. LUCHINS, ABRAHAM S., and LUCHINS, EDITH H. "New Experimental Attempts at Preventing Mechanization in Problem Solving." *The Journal of General Psychology* 42: 279–97; 1950.
17. LUND, S. E. TORSTEN. "Elementary Statistical Analysis for High Schools." *The Science Counselor* 12: 77–78; 1947.
18. MAIER, N. R. F. "An Aspect of Human Reasoning." *British Journal of Psychology* 24: 144–55; 1933.
19. MEEK, RUTH R., and ZECHIEL, A. N. "Functional Mathematics Teaching." *Educational Research Bulletin* 19: 479–82; 1940.
20. MILLER, NORMAN. "More Exercises on Functions." *The School.* Secondary edition 33: 528–29; 1945.

21. MONTGOMERY, G. C. "What Does it Cost to Own and Operate an Automobile?" *The Mathematics Teacher* 35: 15–17; 1942.
22. POINCARE, H. *The Foundations of Science.* (B. Halsted translation). New York: The Science Press, 1929.
23. POLYA, G. *How To Solve It.* Princeton, N. J.: Princeton University Press, 1948.
24. REID, JOHN W. "An Experimental Study of 'Analysis of the Goal' in Problem-Solving." *The Journal of General Psychology* 44: 51–69; 1951.
25. SCHACHT, JOHN F., and KINSELLA, JOHN J. "Dynamic Geometry." *The Mathematics Teacher* 40: 151–57; 1947.
26. THORNDIKE, ROBERT L. "How Children Learn the Principles and Techniques of Problem-Solving." *Learning and Instruction.* Forty-Ninth Yearbook, Part I. National Society for the Study of Education. Chicago: University of Chicago Press, 1950.
27. WESTAWAY, F. W. *Scientific Method.* London: Blackie and Son, Ltd., 1931.
28. WOODRUFF, ASAHEL D. *The Psychology of Teaching.* New York: Longmans, Green and Co., 1951.
29. WOODWORTH, ROBERT S. *Experimental Psychology.* New York: Henry Holt and Co., 1939.

9. Provisions for Individual Differences

ROLLAND R. SMITH

VAST bodies of data testify to the fact that individual pupils of the same age vary widely in respect to a large number of traits. They vary not only in abilities but also interests and needs. There are wide differences in their rates of learning and the degrees of retention. They differ also in such things as emotional, cultural, and economic backgrounds. Even without this experimental data we would know that the differences exist. We do not know so well how to deal with them.

Brief Historical Statement. The fact that individuals vary is so obvious that mention of it goes back many centuries but it has been studied experimentally and subjected to quantitative measurement only in relatively recent years. Plato said that every individual should perform those tasks for which he was best qualified by nature. Aristotle recognized individual differences and the influence of education upon them. The Romans and the educators of the Renaissance also saw that not every one has outstanding gifts or outstanding ability. Rousseau said that each child has his own cast of mind in accordance with which he must be directed. But Francis Galton was the first, about one hundred years ago, to undertake a systematic and statistical study of individual differences.

The first traits considered were comparatively simple ones— memory for nonsense syllables, keenness of eyesight and hearing, color vision, perception of pitch and weight, sensibility to pain, and reaction time to various things. The results of these investigations contributed little to our understanding of the higher and more complex mental processes. Extensive and intensive scientific psychological study of the more complex processes began with the Binet-Simon intelligence scale of 1905.

Binet's greatest contribution was the idea that an individual should be studied through the higher functions rather than the simple sensory-motor processes. The tests he used were administered to individuals. They were followed in 1908 by one revision and again in 1915 by the Stanford revision. In 1916 came tests given to groups—the Army Alpha and Beta intelligence tests— produced by a group of psychologists. Since then hardly any aspect of human variability has gone unexplored. In addition to general intelligence tests there have been tests of personality, of the emotions, of aptitude for general and specific undertakings, and achievement tests in various subject matter fields. The fact of variation in all traits is so well established that it is taken for granted as a problem to be faced by any one who has anything to do with education.

Variation in the Same Grade. The new teacher has a background of these findings of experimental psychology in his or her training for teaching. But he does not have to be in the classroom long to find out for himself the fact of variability. In the first grade Johnny can count to 20 but does not know what he is doing. Mary can get only as far as 10, but she can tell whether she has four pieces of paper or seven. She knows when one group is larger than another. Fred is quite intelligent but he is so shy the teacher has a hard job drawing him out. On the other hand, Jim wants to talk all the time whether he has anything to say or not. Jenny is very proud because she has learned even before she came to school that four apples and three apples are seven apples and two pencils from ten pencils are eight pencils. Hazel thinks she can count but she says, "One, two, three, nine, hundred, million." And so it goes no matter in what grade it is, whether it is in the elementary or secondary school. And the farther along in school it is, the greater the variability. It is not unusual, for example, to find a range of several years among pupils designated as sixth-grade pupils.

A typical situation is seen in the following table showing grade standing in arithmetic of all the pupils in the ninth grades of various schools of a city where the total school population is approximately 21,000. The results were obtained by giving a well known standardized achievement test. In this particular school system, the median grade equivalent for each of the schools is above the expected grade equivalent. But note that there are great variations from the median. For the whole grade, 9 per cent were found to be retarded 1–10 months, 10 per cent 11–20 months, and 3 per cent 21–30 months. In each school, the number retarded up to two years is large enough to provide for in special groupings; the number retarded from 21–30 months is not large enough to think of in this way. There are many retarded. On the other hand, the table shows large numbers advanced beyond the expected norm. The most important fact to note for our purposes is the spread of many years of arithmetic achievement in the one grade. A similar table made for the third grade would show about the same thing, except that the range would be less.

Whether a school be large or small, teachers face the problem of what to do to adjust their teaching to pupils whose abilities and achievements cover a wide range. In a democracy the objective is to give every pupil the best education of which he is capable.

Method of Adjusting to Individual Differences. In most school systems it is agreed that there shall be little if any differentiation of topics in the first eight grades. There is a core of subject matter needed by all normal citizens. Differentiation here will therefore be not in topics but in levels of learning and depth and scope. The amount of concrete background in any topic can be varied. Rates of teaching can be varied. The extent of a topic can be varied. Each pupil will be expected to do work only up to his own capacity to learn. He should master what he does at his own level of maturity so that he can do things at that level efficiently and can have a sense of having achieved success.

Teaching so as to make variations in depth and scope is made easier by ability grouping. When this is done for a class as a whole, it is an administrative matter and the responsibility of the principal. But there is no such thing as homogeneous grouping. If there are only six pupils in a group, no matter how carefully they are selected, there will be six different individuals, with six different sets of abilities, interests, and needs. No matter how efficiently a group is organized, the teacher will have to do a good deal of individual diagnosing and teach accordingly.

Beginning with the ninth grade, there can be variation in courses as well as in depth and scope. There will always be a place for the so-called sequential courses of academic mathematics—

274 THE LEARNING OF MATHEMATICS

GRADE IX—APRIL 1951—ARITHMETIC

	SCHOOL A		SCHOOL B		SCHOOL C		SCHOOL D		SCHOOL E		SCHOOL F		WHOLE GRADE	
	No.	%	No.	%	No.	%	No.	%	No.	%	No.	%	No.	%
Retarded														
1–10 months	33	14.4	18	8.1	10	7.3	28	8.1	6	7.8	10	6.9	105	9.2
11–20 months	32	14.0	24	10.8	20	14.6	20	5.8	4	5.2	14	9.7	114	9.9
21–30 months	12	5.3	10	4.5	8	5.9	5	1.4	2	2.6	0	0.0	37	3.2
31–40 months	1	0.4	2	0.9	1	0.7	2	0.6	0	0.0	0	0.0	6	0.5
41+	0	0.0	0	0.0	0	0.0	0	0.0	0	0.0	0	0.0	0	0.0
At expected grade equivalent	5	2.2	4	1.8	0	0.0	4	1.2	0	0.0	0	0.0	13	1.1
Advanced														
1–10 months	33	14.4	30	13.5	24	17.5	39	11.2	11	14.3	14	9.7	151	13.0
11–20 months	22	9.6	23	10.4	18	13.2	33	9.5	10	13.0	13	9.0	119	10.3
21–30 months	37	16.1	39	17.5	22	16.0	66	19.0	16	20.8	18	12.5	198	17.1
31–40 months	35	15.3	21	9.5	17	12.4	56	16.1	11	14.3	75	52.2	215	18.6
41+	19	8.3	51	23.0	17	12.4	94	27.1	17	22.0	0	0.0	198	17.1
Totals	229	100.0	222	100.0	137	100.0	347	100.0	77	100.0	144	100.0	1156	100.0
Median grade equiv.	10.6		11.6		11.1		12.4		12.2		12.9		11.7	

Expected grade equivalent 9.7.

algebra, geometry, solid geometry, and trigonometry. But there is an increasing number of pupils in our secondary schools who cannot profit by these more or less professional courses. For these pupils, courses in general mathematics have been provided. These courses can be made much more flexible than the academic courses and so can be fitted to a wide range of interests, abilities, and needs.

Diagnostic Testing for Remedial Work. One form which provision for individual differences takes is diagnostic testing with subsequent attention to errors made by individuals. This is done on the assumption that the study of mathematics is the acquiring of particular understandings and skills. This is only narrowly true. Occasionally a child or a student may show particular weaknesses. He may be good in general but make errors in subtraction with borrowing. A little remedial work with that particular thing may well be all he needs. He may have been absent from school when some topic in algebra or geometry was presented and the results of a diagnostic test will show his weaknesses there. The student is very likely aware of the weakness even before the test is given. In general, however, errors in mathematics are not due to this particular kind of thing where repairing by means of patching is effective. If a student is weak in one respect he is likely to be weak all along the line. The real problem is to find out as soon as possible on what level each one of the students can work and teach accordingly. This chapter addresses itself to this point of view.

Bright, Average, and Slow Children. Human beings are not divided into types. There are continuous gradations from one extreme to the other. As is well known, there is a concentration about a central point with the frequency decreasing as the distance from the central point increases. However, for the sake of analysis of learning abilities it is well to think of three groups, the bright, the average, and the slow. It should be understood that there is no well defined line of demarcation between the groups.

Approximately 20 per cent to 25 per cent of the school population at the upper end of the scale may be designated as bright pupils. They have IQ's from 110 up. They find abstract reasoning much easier than do those lower in the scale. They do not need a prolonged period of concrete background. They can more easily grasp several ways of doing a thing, compare them, and choose the best way. A rich associative background helps them with anything they have to recall. Their interest lies in larger units of work with greater returns rather than in short specific assignments. They are critical of their own work. They dislike a large amount of routine. They have a longer attention span. Pupils in this group can carry on a discussion for a considerable period with interest.

At the other end of the scale there are 20 per cent to 25 per cent of the pupils that belong to the slow learning group. With IQ's from 70 to 90, their learning characteristics are different from those described for the bright group even though different only in degree. The difference can easily be seen if we take the same list of characteristics and modify them. They find abstract reasoning difficult rather than easy. They need a prolonged period of concrete background instead of a short one and some of them never leave the concrete level. They learn by simple mental processes and need a large amount of drill and repetition. When they seemingly learn beyond their ability to comprehend, the learning is most often merely verbal memory. Rich associations tend to confuse and bewilder rather than clarify. They are interested in short time units and specific assignments. They have limited powers of self criticism. As one example of this, it is difficult to get them to check even a routine example. They often do not know whether they are right or wrong and seem to care less. Their attention span is short. They tire easily.

In the middle, the members of the average group, 50 per cent to 60 per cent of the total school population, have characteristics between those of the bright and slow groups. They can think in abstract terms provided they have been prepared for it by varied concrete experiences. Properly directed they will check and criticize their own work. They enjoy a certain amount of routine but need to have varied work to keep them interested.

The intelligent quotient is by no means an infallible criterion for dividing children into these groups. There are other characteristics such as initiative and industry to be taken into account.

Many pupils with mild physical defects which may not have been discovered do not achieve as much as their mental abilities would make possible. Habits of work arising from past experience and especially home background may put pupils into a higher or lower class of achievement than their IQ's would indicate. It is not unusual for a boy with an IQ of 100 to do as well in a given subject as a boy with an IQ of 115. At the same time it is not likely that a boy whose IQ puts him in the slow group can compete on equal terms with a boy in the superior group.

Whether a teacher has a selected group—one made up entirely of pupils in one of these categories—or a heterogeneous group where he has to make his own selection, he should take account of the characteristics that have been listed here. He should not deal with slow pupils by the same methods he uses with the superior pupils even if they are in the same class or even if he is developing a topic with both groups at about the same time.

Slow learners, as has been indicated, need more repetition, more concrete development, more trial and error experience, and simpler reasoning. They need help with their reading and vocabulary. They need more help in making generalizations. As a matter of fact, one means of differentiation is to let the better pupils do the more difficult work, make the generalizations, and suggest applications, while the slower pupils are listening. This is the way it works in any heterogeneous group of adults. The slower profit by the work of the more superior.

One problem that always rises when there are fast and slow learners in the same class is what to do with the better pupils while giving extra time to the slower ones. The usual answer is "enrichment" material which may mean different things to different persons. It does not mean giving a set of drill exercises to the brighter students who do not need it. It does not necessarily mean assigning pages of the textbook containing supplementary material, although this is one good solution. It does mean taking advantage of the fact that these brighter pupils enjoy taking units of work that are not routine in nature. Let them work on group projects that require study in the library or in supplementary books provided in the classroom by the teacher. The origin of weights and measures, standard time, arithmetic puzzles, and magic squares are some of the things that come to mind for pupils of fifth- or sixth- grade level. When these groups are ready to make a report, it can be made to advantage to the whole class.

Levels of Learning. When provision for individual differences is mentioned, many of us are likely to think of ability grouping as the basic remedy. It is not the basic remedy. Too often teachers make no distinction in teaching slow groups or bright groups. Grouping is not the remedy when all groups are taught by the same methods. Unless teachers have the concept of various levels of learning and can teach accordingly, grouping is of no great value.

An illustration will show what is meant by the term levels of learning. A child who can find the sum of 5 apples and 3 apples when the apples are present is working one level of learning. Another who can find the sum of 5 apples and 3 apples by drawing circles is working on another level. A child who knows automatically that 5 and 3 are 8 is working on still another level—this time on a mature level.

The concept of levels of learning recognizes the fact that children of various degrees of ability or maturity can learn to do the same thing on several different levels from the simple concrete to the more complex abstract. A good many teachers now in service began teaching under a psychology that demanded the teaching of all processes on a mature level. "Teach a process in the way it is to be used," was the slogan. A pupil's introduction to a process was as an adult would use it. A newer psychology has us accept at first the crude attempts of the pupils and then refine the attempts as the pupil gains more experience.

The methods of dealing with individual differences on the basis of various levels of learning have received more attention in the elementary school than in either the junior or senior secondary school. For this reason, a large number of illustrations have been taken from this part of the school system. Secondary-school teachers, as well as elementary-school teachers, will find a study of these illustrations profitable and can adapt the methods to their own problems.

Suppose a teacher of an unselected group of third-grade children is attempting to develop the first fundamental ideas of division. This is the problem she has given. "Frank has 17 cents. How many 3-cent stamps can he buy?" The members of this class have had two years' experience with numbers. They have learned that they can solve problems by counting things and by grouping things. Many of them know they can use counters such as milk bottle tops in place of actual things. They have often solved problems by drawing pictures or using symbols such as small circles. Some of them can count by 2's and 3's. Now they are all faced with a new problem. How should the teacher handle the situation?

The teacher should at first give no suggestions but watch to see what is being done. Some may solve the problem by using counters. Others may draw circles. Still others may count back from 17, making a mark after every three numbers. It is possible that a few of the children may know that five 3's are 15 and six 3's are 18 and so know almost automatically that the answer is 5.

If the members of the class fail to respond readily, the teacher may make suggestions on the various levels letting each child choose the way that appeals to him. It is better to make brief suggestions than it is to show methods step by step. Over a period of two or three days, these are some of the ways the example can be done.

1. Take 17 counters and place them in piles by 3's. There will be 5 piles and 2 counters left over, showing that Frank can buy 5 marbles and will have 2 cents left.

2. Draw 17 small circles and mark them off by 3's.

3. Start counting backwards from 17. Say 17, 16, 15, and put a mark on paper. Then say 14, 13, 12 and put another mark on paper. And so on.

4. Make a table as follows:

No. of marbles	1	2	3	4	5	6	7
No. of cents	3	6	9	?	?	?	?

5. Start with 17, subtract 3. Subtract 3 from the remainder,

and 3 from the new remainder, and so on. You will subtract 3 five times and have 2 cents left over.

```
   17
    6
   --
   11      2
    6      2
   --
    5      1
    3      5
   --      --
    2
```

6. Buy 2 marbles at a time. Two marbles will cost 6 cents. You can buy 2 and 2 and 1 and will have 2 cents left. (See box above).

7. Buy 4 marbles. Then you will have enough left to buy one more.

8. Buy the largest number of marbles possible all at once. Those who know the multiplication table of 3's can do the example this way.

After a discussion of these various methods of doing the problem, another of similar nature should be given; for example, "How many apples at 4 cents each can I buy for 25 cents?"

Now is the time to watch carefully to see what level each child chooses. You may find some who can get the answer automatically. Some will subtract 4's. A good many are likely to do it objectively. Some will not have grasped the idea at all.

It is a good thing now to divide the class into groups. Make one group of those who need help to do the example even objectively. Make another group of those who can do it objectively, and another of those who are doing it abstractly.

While you are working with the slowest group, let the members of the other groups discuss their methods. There will be different levels of learning even within the groups. In the second group there will be various objective methods. Let the children decide which is the quickest method. In the most mature group there will be some who are subtracting 4's singly, some who are subtracting various multiples of 4, and a few, perhaps, who are doing the whole thing at one stroke. Let the members of this group agree upon the best method. Here will be a strong motivation for learning the multiplication facts.

The teacher's responsibility is to help the slowest group to learn how to do the example objectively and then as they learn better how to subtract 4's, to do it by subtraction. They should go on to still higher levels as and if they become ready. All the children should have a chance to discuss the various methods and choose the best method. Those on a lower level of learning should not be forced to go higher but should be encouraged to do so when they can.

This method of procedure is in marked contrast to the method which does not let a pupil progress to a new topic until he has grasped the preceding one from the mature point of view. This method allows progress even though the level of learning is not a mature one. This method does not keep a boy from learning what is meant by division, or how to do simple exercises in division, until he has mastered the multiplication facts. He can do division by pictures. He can do it by subtraction. He may need to have a multiplication table before him. At any rate he can progress.

We now take another topic illustrating teaching on various levels of learning, this time from the topic of fractions in the fifth grade. Since the pattern of procedure is much the same as the one given, we shall merely outline it.

Suppose it is desired to develop the topic of subtracting a fraction from a whole number. The procedure might be as follows:

The first problem is a concrete one—"I have 6 feet of rope and cut off half a foot. How much do I have left?" The pupils have no mature method for doing this example. It has not been taught. Ask for an answer found by any method at the pupils command. If you get answers, ask how the answers were found and discuss the methods. If answers are not forthcoming suggest (a) counting by halves—$\frac{1}{2}$, 1, $1\frac{1}{2}$, $2\frac{1}{2}$,—or (b) drawing 6 circles and dividing one of them into 2 equal parts. Crossing out one-half of one of the circles will give the answer.

Now try a few other examples involving the subtraction of such fractions as $\frac{1}{4}$, $\frac{3}{4}$, $\frac{1}{3}$, $\frac{2}{3}$, $\frac{1}{6}$, $\frac{5}{6}$ and let your pupils find the answers by their own methods. Discourage pencil computations.

At this stage few if any pupils will need to bring a rope to class and carry through the operation thus objectively. Most of them will do it by drawing circles and dividing one of them into fourths, thirds, or sixths, as the case may be.

$$6 = 5\tfrac{?}{4}$$
$$\tfrac{3}{4} = \tfrac{3}{4}$$

Now suggest that these examples can be done without the use of pictures. Put an example like the one above on the board. There is just enough help here to permit the brighter ones in the class to carry on. Let the others continue objectively. Do not require any pupil to do it abstractly until he is ready to do it that way. And when he does it abstractly he should be asked to explain his work and to show the very obvious connection between drawing six circles and dividing one of them into fourths and the fact that $6 = 5\tfrac{4}{4}$.

This work has taken probably not more than twenty minutes. An objective method has been shown for the class as a whole and understood by most of the pupils. The better students have graduated quickly to an abstract method. They have associated the objective and the abstract quickly.

Now, just as in the example of the teaching of division, it is time to divide the class into groups, helping the slow group to see what it is all about, giving more practice to the middle group and helping those in it to associate the steps in the abstract method with the corresponding steps in the concrete.

The bright group, because superior children are helped by rich associations, may well learn that this subtraction is only a special case of a larger pattern of so-called borrowing. You can show the following examples.

1. "From 8 yd. take 2 ft." This is done by changing 8 yd. to 7 yd. 3 ft.

2. "From 9 wk. take 6 da." This is done by changing 9 wk. to 8 wk. 7 da.

3. "From 8 dimes take 6 cents." This is done by changing 8 dimes to 7 dimes 10 cents.

PROVISIONS FOR INDIVIDUAL DIFFERENCES 283

4. "From 80 take 7." This is done by changing the 8 tens to 7 tens 10 ones.

The pattern is obvious. It is identical to the one that has been used in subtracting a fraction from a whole number. This kind of generalization is better done with bright pupils than with poor pupils. Too much of this can be more confusing than helpful to slow pupils.

These two illustrations have been taken from topics in the elementary school. The same fundamental ideas apply to all grades, in the secondary school as well as the elementary school. The following illustration is taken from algebra.

Verbal Problems in Algebra. Most pupils have difficulty with verbal problems. Some pupils have more difficulty than others. Teachers find the pupils on many levels of attainment. Many cannot read with comprehension. Others do not readily see the relationships involved. There is a definite need for developing the procedure in problem-solving objectively so that a distinction may be made for different levels of maturity.

Objectivity in this case does not mean going back to the handling of things as it does in the early years of the elementary school. It means going back only to abstractions already well established. Consider this problem for example.

"A collection of nickels, dimes, and quarters amounts to $4. There are 10 more nickels than dimes and 2 less quarters than dimes. Find the number of each."

It is assumed that all members of the class can do the following problem: "What is the total value of 12 dimes, 22 nickels, and 10 quarters?" It is assumed also that they can do this problem: "I have 10 more nickels than dimes and 2 less quarters than dimes. If I have 12 dimes, how many nickels and how many quarters do I have?" Any pupil who cannot do these two problems, involving arithmetic numbers only, has a poor chance of being successful in algebra. This work with numerical relationships is a sufficient basis for objectivity.

When a class is first confronted with the algebraic problem stated above, the teacher should ask, "Suppose there are 12 dimes, how many nickels and how many quarters would there be?" This question forces reading with comprehension and shows the teacher those who need help in that direction. Pupils need to learn that there is a relationship given between the numbers. The next question would be, "What is the total value of these coins?" The final question is, "Is this the total given in the problem?" It is not the right total, so we know that 12 dimes is not the right number.

Before going through this problem again with some other assumed answer, not 12, it is well for the teacher to go on to the algebraic solution for the benefit of the brighter pupils who are ready for it. The algebraic solution should be associated step by step with the numerical work just finished. Briefly the discussion would be as follows.

"We know that 12 dimes is not right, but we have discovered and used all the relationships stated and assumed in the problem. Instead of 12, let us use n. There are n dimes. How did you get the 22 nickels in your previous work?" *Answer:* "I added 10 to the number of dimes." "What is the number of dimes in our present analysis?" *Answer:* "n". "What is 10 added to n?" *Answer:* "$n + 10$." "There are $(n + 10)$ nickels. How did you get the number 10 for the quarters in your previous work?" *Answer:* "I subtracted 2 from the number of dimes." "There are $(n - 2)$ quarters." "How did you find the value of these coins?" *Answer:* "I multiplied the number of dimes by 10, the number of nickels by 5 and the number of quarters by 25." "What values do you get for n dimes, $(n + 10)$ nickels, and $(n - 2)$ quarters?" *Answer:* "$10n$ cents, $5(n + 10)$ cents, and $25(n - 2)$ cents." "How did you check before to see if 12 dimes was correct?" *Answer:* "I found the total value to see if it was $4. "You will do the same thing here. What will you have when you say that the sum of the values must be $4?" *Answer:* $10n + 5(n + 10) + 25(n - 2) = 400$.

This is an exceedingly brief statement of the discussion. In actual practice the reactions of the pupils would be taken into account and the discussion would be less formal and better directed.

You now have the members of your class thinking on various levels. Some will be able to think in terms of an algebraic unknown, others will still be quite confused. In between, there will be those who have begun to see what it is all about, but need much more practice in discovering the relationships by means of an assumed arithmetic answer. When you develop your work on various levels, you have a better chance of interesting all the members of a class. You avoid the hopeless confusion that often besets the slow pupil and you challenge the brighter pupils.

This matter of levels of learning is fundamental in providing for individual differences whether the provision is made on an individual basis or in groups. Some pupils may reach a certain level of development and stop there. They may have reached the limits of their ability. The difficulty, however, may be something else. It may be some emotional maladjustment or faulty teaching. It is not wise for a teacher to assume that the limit of capacity has been reached. It is not likely that any dull child will become a genius. At the same time, there is always hope that the right surroundings, the right conditions, and a more adaptable form of teaching may so appeal as to raise a child beyond what his past record would predict.

Teachers should keep constantly in mind the fact that learning involves the learner. Unless the learner can be guided to think, feel, and act appropriately in a given situation, it is not possible for him to learn what is intended. The teacher who attempts to teach always on the mature level, merely showing pupils examples of work properly done with little or no attempt to accept and develop the pupil's first crude methods, is not likely to find his pupils interested enough to enter into the work wholeheartedly. Mechanical drill on top of this kind of teaching is doomed to failure. When pupils find things to do on their own level of learning, they are most likely to enter into the situation.

Differentiation in Depth and Scope. The mathematical program for Grades I through VIII should be essentially the same for all normal pupils. There is no question about this for Grades I through VI. The same general statement applies to Grades VII and VIII. We need to make certain that the essentials for functional competence are achieved by all who can learn them and the seventh and eighth grades are crucial years in the attainment of that objective. The pupils in these grades, however, cannot all learn the same amounts nor at the same rates. To take care of the differences, there must be differentiation but it should be in depth and scope not in topics. We have already discussed varying levels of learning. Following is a brief discussion of differentiation in depth and scope.

The term "depth and scope" refers not so much to levels of learning as it does to graduation of subject matter within topics. To be sure, some topics cannot be differentiated. The multiplication table, for instance, is the same for all pupils. However, most topics can be differentiated. In fractions, all children should understand the meaning and the processes of adding, subtracting, multiplying and dividing even though it is done only on the picture level. Almost all the children can work with the simplest kind of examples. The faster pupils should progress more rapidly to more difficult examples and more difficult applications. In decimals, too, the meaning and elementary skills should be developed in all four processes, but the extent to which any child or class should go depends upon the ease with which they grasp what is being done. In per cents let all the pupils have practice in the meaning and work with the simplest examples of per cent— 50 per cent, 25 per cent, $33\frac{1}{3}$ per cent, and so on—per cents that can be treated as fractions. Those who can should go on to work with other per cents such as 56 per cent, $73\frac{1}{2}$ per cent, 4.5 per cent, $\frac{1}{2}$ per cent.

Teaching by Wholes. Giving first a birds' eye view of a whole topic on a small scale is an excellent means of providing for individual differences. In most seventh and eighth grades there are several topics required for a year's work. Ordinarily a tentative time schedule is given as a teachers' guide. But teachers, more often than not, complain that they need more time for teaching each topic. They should try teaching the essential parts of a whole topic in the first few days assigned to that topic. In superior classes this can be done in a very short time and the remaining time left for the development of details and for enrichment. In poorer classes this preview of the whole will take longer. Whatever time is left can be used for the important details that are left. However, since the most essential parts of the topic were completed early in the schedule, the teacher will feel no qualms of conscience in progressing to the next topic when it is time to do so.

Following is a brief report of three days' work in decimals taught from this point of view. Note that it begins with the meaning of decimals and goes through division by a decimal. It gives the whole story but only on a small scale. Only tenths are discussed, not decimals in general.

For the first day:

1. A brief background of fractions. If a line is divided into 2, 3, 4, or 5 equal parts what is one part called? Count by halves, thirds, fourths.

2. A line 2 feet long with each foot divided into ten equal parts. One part is 1/? of a foot. Count by tenths pointing to the corresponding marks on the scale:

$$\tfrac{1}{10}, \tfrac{2}{10}, \cdots, \tfrac{9}{10}, 1, 1\tfrac{1}{10}, \cdots, \cdots, 2.$$

3. Another way of writing tenths:

$$\tfrac{1}{10}, \tfrac{2}{10}, \cdots, \tfrac{5}{10}, \cdots, \quad 1$$

$$.1, \ .2, \cdots, \ .5, \cdots, 1.0$$

$\tfrac{5}{10}$ is $\tfrac{1}{2}$ so $\tfrac{1}{2}$ and .5 have the same value. Point to the .5 mark on the line. This is $\tfrac{1}{2}$ inch.

4. For meaning. Read 2.3. Write it using a common fraction. Are 3.4 pies as much as $3\tfrac{1}{2}$ pies? Where is 1.3 on the scale? How does it compare with $1\tfrac{1}{2}$? What does 2.5 mean to you? ($2\tfrac{1}{2}$).

5. About how much is 2.1 × 3.2? (It is about the same as 2 × 3.). $4.3\overline{)16.2}$ is about how much? (About the same as $4\overline{)16}$.) 2.4 + 3.1 is a little more than __?__ .

All the work of this day was pointed toward meaning.

For the second day:

1. Review of first day's work. Which is larger, 2.3 or 23? If you had .3 pound of candy, how much less than a half pound would you have? When you see 12.5, what do you think of? ($12\tfrac{1}{2}$). Count from 7 to 8 by tenths and write the numbers using a decimal point. Count from 0 to 1 by tenths pointing to the corresponding marks on the scale. Count by tenths from 7.5 and write

the numbers until you have written seven numbers. Between two what whole numbers is 4.7?

2. Addition. Note how much easier it is to work with decimals than with fractions. Addition without carrying:

$$6\tfrac{3}{7} \qquad 4\tfrac{3}{10} \qquad 4.3$$
$$5\tfrac{2}{7} \qquad 5\tfrac{4}{10} \qquad 5.4$$

Addition with carrying:

$$5\tfrac{4}{8} \qquad 7\tfrac{6}{7} \qquad 8\tfrac{9}{10} \qquad 8.9$$
$$7\tfrac{5}{8} \qquad 4\tfrac{3}{7} \qquad 4\tfrac{7}{10} \qquad 4.7$$

3. Subtraction.

$$4\tfrac{1}{7} \qquad 7\tfrac{3}{10} \qquad 7.3$$
$$2\tfrac{3}{7} \qquad 4\tfrac{7}{10} \qquad 4.7$$

We can do this work just as if there were no decimal points and then put the point in the correct place in the answer. Be sure to get the points in a column.

4. For meaning. Give estimate only: $3\tfrac{3}{10} \times 5$. Also 3.2×5.

For the third day:

1. Written review. Write as decimals $2\tfrac{3}{10}$, $\tfrac{1}{2}$, $7\tfrac{1}{2}$. Which is larger 4.8 or 32? Is 3 less than, equal to, or greater than 3.0? Is 4.7 more or less than $4\tfrac{1}{2}$? How much? Add 6.2 and 3.4; 5.7 and 6.8. Subtract 2.4 from 6.2. Subtract 3.7 from 5. Is 3.2×2.1 near to 1, 6, or 144? Is $16.3 \div 2.1$ near 4, 8, 16, 50 or 144?

2. Multiplication. Estimate $3\tfrac{1}{2} \times 4$. Also 3.5×4. Find the exact answer to $3\tfrac{1}{2} \times 4$. Multiply 3.5 by 4 paying no attention to the decimal point. Place the decimal point in the answer according to estimate. How do your two answers compare?

Practice with several of like nature.

Then 12.2×12, placing the decimal point by estimate. Avoid examples like $64 \times .8$ because of difficulty in estimating.

3. Division. Assume the ability to multiply by 10 previously taught. What is the quotient in each case?

$$2\overline{)8}, \qquad 4\overline{)16}, \qquad 8\overline{)32}, \qquad 20\overline{)80}$$

(To develop the fact that multiplying the divisor and the dividend by the same number does not change the quotient.)

What is an approximate answer to 2.1)̄105 [50]. This is difficult because the divisor is a decimal. By what could you multiply 2.1 to make it a whole number? [10] Then if you multiply the dividend also by 10 the quotient will be the same as in the original example. The examples become 21)̄1050.

Practice. The final idea is this: Never divide by a decimal. Change it to a whole number.

To be sure, skills have not been developed in this short time. But the big ideas have all been presented. From now on the students have enough background to proceed at their own rate. The details will be more meaningful because they will be fitted into the whole picture.

Another illustration of giving a preview of a topic at the beginning, this time per cents. The important subtopics are: meaning of per cents, finding a per cent of a number, finding what per cent it is, discount, commission, per cent of increase and decrease, and finding the whole when a part is given. One way to proceed would be to take each subtopic and treat it at length. Then in some classes there would be the question of having time to complete all the subtopics. A better way is to discuss all these subtopics using only 50 per cent, 25 per cent, and 75 per cent. Since these per cents can be easily converted into simple usable fractions, a big overview of the whole topic can be given in a short time. The remaining time, varying in length according to the class, can be used for developing further details. The following exercises are suggestive of the procedure.

1. Find 50 per cent of the following numbers: 16, 52, 100, 432, ½, ⅓, ⅔, $2.50.
2. Find 25 per cent of the following numbers: 24, 60, 72, 234, ½, ⅓, ⅘, $6.00.
3. Find 75 per cent of the following numbers: 24, 60, 72, ½, ⅓, ⅔, $6.00.
4. John had 75 per cent of his examples correct. There were 12 examples in all. How many examples did he have correct?
5. Draw a rectangle and shade 50 per cent of it.

6. I saw a suit marked $60. A friend of mine owned the store and he told me he would sell the suit to me at 25 per cent off (discount). What will I have to pay for the suit?

7. I sold $40 worth of Christmas cards for the holiday and was allowed to keep 25 per cent of that amount for my work (commission). How much could I keep?

8. Frank is making $40 a week this year. Next year he will get a 25 per cent increase. How much will he then get a week?

9. A certain town used to have a population of 200 persons. It has decreased 25 per cent. What is its population now?

10. The Blues played 8 games and won 4 of them. What per cent of the games did they win? (Think: 4 out of 8 is what fraction?)

11. A pint is what per cent of a quart?

12. I bought a top for 12 cents and sold it for 15 cents. What was the amount of increase? What was the per cent of increase?

13. A $4.00 doll, if sold at a 25 per cent discount, will sell for __?__ .

14. During a test of 20 examples, the teacher said, You must get at least 75 per cent of these examples right before you can go on to the next topic. How many examples was that?

15. If 25 per cent of a number is 8, what is the number?

Grouping. Homogeneity of ability in a group is impossible. No matter how carefully a group is selected there will be differences. The range of ability, however, can be reduced. When the range is too great it is difficult to teach through group instruction. The purpose of grouping in ability is to make it easier for the teacher to put into practice what has been said here about levels of learning and differentiation in depth and scope.

Grouping is of two main kinds, grouping within a class and grouping into classes from a larger population. We have already mentioned grouping within a class in connection with various levels of learning. We shall return to grouping within a class later in this section. Now we should like to make a brief statement concerning grouping into classes.

Ability grouping by classes is particularly valuable in the seventh and eighth grades where all pupils are expected to deal with the same topics. In this case grouping is not the responsibility of the teacher but of the principal. It cannot be done unless the school is large enough to make at least two classes in the same grade.

Attempts have been made to divide a grade into as many classifications as there are possible classes. For example, when the numbers are large enough for six classes, the pupils would be divided into six groups according to ability. This has not worked too well in the lowest group. The lack of industry, interest, and responsibility among those who make up a large percentage of the low group make them hard to handle when they are all together. Better results are obtained when a total population is divided into two groups according to ability and classes formed from these two groups. Then the range is not so great as in a non-selected group and some of the disciplinary troubles of the low group are avoided.

We have already said that the IQ is not infallible as a criterion for making subdivisions. It can be used to make a good start but there are always doubtful cases that must come under the scrutiny of principals, counsellors, and teachers on a more individual basis. Past achievement and teachers estimates must be taken into consideration as well as the IQ. Besides this, if it is at all possible, and it is possible if the teacher has the point of view of differentiation in depth and scope and so keeps somewhere near to a time schedule, pupils should be changed from class to class when it becomes clear that he is maladjusted.

Again we wish to make emphatic that grouping alone does not take care of individual differences. What is done within the groups is what counts.

We return now to a report of a lesson in a third grade where very definitely the teaching took into consideration the differences in ability of three groups. This was in January. The class was organized in September.

The class had been divided and taught since October in three groups selected according to reading and arithmetic ability. The groups were flexible. Children had been changed from group to group as the necessity demanded. The teacher said she could well have had 6 or 10 groups so far as the range of abilities was concerned but three groups were all she could handle efficiently. On this day Group I contained 14 children, Group II had 9, and Group III had 9.

The assignment for the three groups was on the blackboard. Three different textbooks were used, mainly because of the difference in vocabulary required. This is what was written on the board (the names of the textbooks are not mentioned).

Group I. Name of textbook, page 201, rows 1 and 2.
Group II. Name of textbook, page 36, rows 5, 6, 7.
Group III. Name of textbook, page 120, set 1.

For all groups, (to be done, when and if the other assignment was finished):

$$\begin{array}{cc} 6 \text{ balls} & 12 \text{ tops} \\ +8 \text{ balls} & -6 \text{ tops} \\ \hline \end{array}$$

Make pictures to prove them.

If you have 12 books and have read 2 of them, how many do you have left to read?

26 means _____ tens and _____ ones.

The exercises in the assignments were exercises which the teacher was reasonably sure the children could do. The purpose was to establish knowledge and skill already developed. No questions could be asked by the members of a group doing the exercises because the teacher was busy with another group. The three assignments were different gradations of the same topic. Group I had addition of dollars and cents, Group II had addition of columns of single digits and addition and subtraction of two numbers of two digits each. Group III had addition of single digits in columns of three and four digits.

Each group was called to the front of the room in succession standing about the teacher. For about three minutes each the teacher checked to see that the assignments were understood. In the first group she had the children find the page and read the directions. She asked how to write 1 dollar and 50 cents, 50 cents, and 5 cents, and how to place these numbers for addition. Individual pupils wrote the answers on the board. Since the exercises in the book had the numbers written horizontally, it was emphasized that the decimal points should be in a column.

The members of Group II were asked about their assignment in somewhat greater detail. "Find the page." "Is 136 more or less than 100?" "What do the directions say?" "Do the first one on the board." "How many tens and how many ones are there in 79?" "Be sure to put the tens under the tens and the ones under the ones." These are some of the statements made and questions asked by the teacher.

Group III had to be questioned carefully about the directions in the book. It was obvious that there were reading difficulties. The discussion went along on a lower level of learning. The column addition of single digits was done by means of beads on a wire. The children were reminded how to do the work by means of small circles made on paper. They were advised to do their assignment by means of circles if they needed to do so.

When all were set to work the teacher went about the room for about three minutes to see if all were working and then she called back the first group for a lesson. This time the children in the group were seated in an arc of a circle in the front of the room with a blackboard handily in front of them.

They discussed first a practical problem arising from one of their social activities. The treasurer of the class had been collecting 25 cents from each member for a project they had decided upon. He already had $7.50 which he counted for the members of the group,—6 ones, 4 quarters, 1 quarter and 5 nickels. He commented on the fact that 4 quarters make a dollar and five nickels make 25 cents. A "quarter" and "twenty-five cents" have the same value.

There were three more children who had promised to bring a quarter each. The teacher asked the question, "How much will you have then?" All the members of the group did the work individually with pencils and paper. One boy said, "It's easy if you know how to count by 25's". Two different methods were shown. About half the children had done the exercise one way and half the other way. One group had added 75 and the other had added $25 + 25 + 25$. Both groups, of course, got the same total. This was because "75 and three 25's is the same thing." They decided that the easier way was to add 75.

The work taken in the next 10 minutes was a consolidation of what they already knew about the table of 2's with the introduction of the times sign, the formal language of multiplication, and the algorism 5 × 2 and 2 × 5. Objective work was intermingled with the abstract. Discussion was intermingled with drill. A beginning was made with the concept of division. "How many balls at 2¢ each can you get for 12¢?" was asked along with such questions as, "How much would 8 balls cost?"

Then the group was returned to its written assignment and Group III was called up for a lesson.

The purpose of the work with Group III was to make clearer the meaning of *one-half*. The teacher tore off a corner of a paper and asked if it was half the paper. The children replied, "No, it is not even." They divided circles, squares and rectangles into two equal parts by estimate and shaded *one-half* of each. Then they turned to getting half of numbers. Each member had 20 small beads strung on a stiff wire and found half of several numbers by trial. They could get ½ of 4, ½ of 8, and ½ of 12, by making sure that there was the same number in each of two groups. Finally the method of writing *one-half* in figures was given and the meaning was explained, 1 of 2 equal parts.

The entire lesson period was about 45 minutes. Group II had a written assignment on this day but did not have a meeting as a learning group. The teacher found it rather impossible to take care of more than two groups a day except for the written assignment.

This description of teaching by groups has been given thus in detail as a suggestion as to how it can be done. Many teachers state that they would like to take care of differences in ability by groups within their classes but do not see how to manage the various groups.

Geometry. In order to think with any clarity concerning provision for individual differences in demonstrative geometry it is necessary first to consider the objectives of the course. As long as we insist that the aim is to have students learn how to prove theorem after theorem and be able to reproduce the proofs at will, there is little hope of a satisfactory differentiation on different levels of ability. There are so many details to think about in such a case that there is no time for the kind of development that is necessary with the slower groups. However, if the aim is to meet a few large objectives as has been suggested of late, much can be done.

There are first of all the facts of geometry, the geometric concepts and geometric statements of relationship. These are fundamental to all students taking geometry. There is also the demonstrative side of geometry. Students should learn what a deductive proof is and how it differs from an induction. They need to learn to distinguish between hypothesis and conclusion and see the relationship of the hypothesis to the whole proof. Before this they must learn the significance of the "If-Then" relationship. They have to learn the futility of attempting to use any statement as an authority before the conditions of that statement are fulfilled. These and a very few others are the larger objectives we should seek in demonstrative geometry.

If we have these larger objectives in mind, differentiation for various groups is easily possible. In schools that are large enough to have separate classes of slower and faster students, courses can be quite different and yet have the same general aims. The slower students will need much more experience with drawing and measuring and with inductive approaches to theorems. The introduction to proof will be much more gradual and informal. Many of the theorems heretofore proved at the beginning of the course should be taken intuitively or after measurement. Congruence theorems can be taken for granted after construction. The fact that the base angles of an isosceles triangle are equal may well be assumed and the fact checked by measurement. A beginning of deduction can be made quite informally by asking about certain figures, "If you measure this angle and find it to be 70°, how many degrees are there in these other angles? Give your answer without measuring." This in connection with such figures as one line meeting another, two intersecting lines, or two parallels cut by a transversal. The theorem about the sum of the angles of a triangle may well be assumed without proof and a variety of corollaries deduced from it. In connection with other topics such as angles and arcs, similar triangles, and areas there will be much measuring, but also informal deductions leading gradually to more formal work.

The course certainly need not be made a travesty on demonstrative geometry but the main objectives can be led up to gradually by means of a laboratory procedure and deduction of an informal sort. The objectives will be met but upon a lower level of learning.

A class of the faster students can proceed with much more abstract work from the beginning. However, a teacher who has taught the course for the slower students will find many opportunities for making the work clearer by a brief use of the laboratory method.

In the small high school where no selection into classes is possible and all levels of students are in one group, the concept of teaching on various levels of learning is still possible by grouping within the class. The problem here is to keep the better students properly occupied while the slower ones are carrying on their experiential work. This is the time to think of "enrichment",— history of theorems and of men behind the theorems, various practical applications of the theorems, the logic of magazine and newspaper articles, fallacies in advertisements and elsewhere. The brighter group will enjoy these things and can report to the whole class about them.

Differentiation in Courses. For years after the great influx of students into the secondary schools, leaders in the field of mathematical education attempted to vary the traditional academic courses in depth and scope to meet the new conditions. Some progress was made, but the continuing large percentage of failures in these courses attested to the fact that they were not meeting the needs of a great number of students. There were two alternatives, (a) those who could not profit by algebra, geometry, and trigonometry should drop out of mathematics, or (b) there should be different courses for these persons.

At first there was a trend toward the first alternative. The drop in enrollment in mathematics became alarming to those who saw the need for mathematics among those who were no longer taking the subject. The war pointed up this need and showed the weakness of our young people in mathematics. From that time, the second alternative has become popular. To be sure, courses in general mathematics go back further than the war, a quarter of a century at least, but the war and the subsequent Second Report of the Postwar Commission of the National Council of Teachers of Mathematics gave impetus to them.

General mathematics courses have a definite place for the review of arithmetic and go on from there to various units of greater immediate value than the algebra and geometry of academic courses. Among other topics, these courses contain the mathematics of the home, the mathematics for citizenship, and mathematics for reading newspapers and magazines. They appeal to students of different interests and needs. They are much more flexible and can be much more easily adapted to different backgrounds and levels of ability.

General Mathematics in the Ninth Grade. One group of students in the ninth grade needs algebra as a basis for further study toward scientific or engineering careers. There are others who may wish to elect algebra because they are interested in mathematics and have done well previously even though they are not thinking of using it as a professional background. The rest, and it is a large group, need a more flexible course in mathematics to meet varying needs.

The problem of dividing students into two groups, one taking algebra and the other general mathematics is not so great as the problem of varying the general mathematics course to meet the needs and the interests of those taking this course. There are prognostic tests to provide a first approximation for a division into the algebra and general mathematics groups. Cooperation between teachers and counsellors provided with all available data concerning the students will make a workable division possible. The real danger is that those who are taking general mathematics will all have to take a set course without much thought given to the wide range of abilities in this group. Students in general mathematics vary not only from year to year but from school to school, from class to class within a school, and within classes. No two classes are likely to need the same kind of material.

We suggest the following very flexible course as a basis for general mathematics in the ninth grade. It has virtue in that it has a diagnostic approach taking into account individual and group differences.

A course in general mathematics in the ninth grade should have a core of arithmetic because, for the most part, arithmetic is the most useful branch of mathematics. Besides, many if not most, of the students enrolled in the course show a decided need for more efficiency in arithmetic. But the course should not be limited to arithmetic.

Our students live in a world of geometry. Geometric figures are all about them. An elementary knowledge of geometric figures and the relationships among them will help the students not only to a greater appreciation of their surroundings but are desirable and often necessary in the kitchen, the shop, factory, and the office. Informal geometry as well as arithmetic should be a part of the course for all students of general mathematics in the ninth grade.

There should be a thread of arithmetic and geometry running through the year's work. Beyond these two fields it is not only difficult but unwise to state specifically what should be studied in any given class. Individual and group differences must be provided for by allowing the teacher and the class to select units from a suggested list of units.

The arithmetic should be approached from a diagnostic point of view. Seldom will a class need the same kind of approach as was given in the seventh and eighth grades. It should be taught as needed, not as new work.

The order of development of details in a topic in the seventh and eighth grades is important. Usually there is enough background in the ninth grade so that order of procedure is not paramount. Often you will learn more about the members of a class by jumping into the middle of a topic or even asking questions that might in an earlier year come at the end of a topic. Instead of taking a subject like division of whole numbers or operations with fractions and giving a careful development from start to finish just as if the work had not been done before, follow the technique of giving short tests of miscellaneous examples several times a week and govern your instruction accordingly. This method of procedure will help you to vary your arithmetic according to the needs of the individuals in your class.

If you are careful to see that your series of miscellaneous tests covers all the essentials in arithmetic, these tests will constitute your course of study so far as the arithmetic is concerned.

Study pupil's errors. Try to find out by careful questioning the causes of the errors. You will find some errors common to many pupils. Other errors will be made by only a few. Plan developments in the hope that with proper understanding many of the errors will not recur. Help individuals as found necessary. Continue to study the reactions of your pupils to everything that goes on in the classroom so that you can check your methods of teaching. This is the kind of thoughtful experience that results in increased ability to make adequate diagnosis of the different needs of your pupils.

Some days this work in arithmetic may take the whole period. More often it should not take more than 20 minutes for the test, the checking by the students, and the discussion. You need not cover all the errors on any one test. Choose the most common errors. Fit the instruction to the time at your disposal. Long periods of specific drill bring quick results but seldom are they lasting. Another day is coming.

Beyond this arithmetic and small amount of geometry, it is difficult to prescribe required units. It is wiser to have a list of elective units. Teachers and pupils together can choose units to meet their needs and interests.

The units in various textbooks on general mathematics give us a good start for a list. No one textbook is enough, however, to take care of the wide range of interests. The check list of units given in the Second Report of the Postwar Commission is another good source. There are more units there than can be studied by any one class.

The units should cover areas of mathematics, such as those in the Postwar list, but they should cover also areas of living. In other words there should be social units as well as mathematical units. The list below is suggestive but by no means exhaustive.

The mathematics found in reading a newspaper.
How I earn and spend my money.
Preparing to go to camp.
The mathematics of a railroad timetable.

Buying a dress. Making an end table.

In dealing with these units, the most important thing is to interest the pupils, to sell them the value of arithmetic in their everyday living. Competence will not come unless interest is aroused. When you have the interest of your pupils, learning can take place.

Many of the students in a general mathematics class will be average or slow learners. Slow children cannot carry on enterprises, investigations, or discussions as complex and as comprehensive as those carried on by brighter ones. They do not see so far ahead, consider so many alternatives, or take into account as wide a variety of factors. Discussion periods must be short. The interest of some classes cannot be held for more than ten minutes at a time on any one thing.

Classes will vary in interest and attention span. A teacher was able to keep one class interested in a newspaper unit for a half hour at a time. The same teacher found another class would stay with him on the same material for only a little over ten minutes. He did not try to keep the second class going for half an hour just because he could do it with the first class. An attempt at free discussion is worth little when interest lags. If you can find units that parallel the life interests of your pupils, you will be able to keep up interest for longer periods of time.

You will often have to carry the load of a discussion period yourself. Pupils will fall in with your suggestions and make suggestions following the pattern of yours but in general you cannot depend upon a great deal of originality. When suggestions are not forthcoming you will have to tell your pupils what to do. Never let a discussion drag or become lifeless because you are trying to get something from your students that is not there. Always strive for the best kind of pupil participation but remember that when interest fails you are probably attempting the impossible.

An illustration will help to make clear the meaning of the preceding paragraph. One teacher carried through a unit on Buying and Maintaining a Home. The unit took fifteen to twenty minutes a day for three weeks. The pupils kept note books in which they recorded important findings and the computations that went with them. The teacher had source material at hand but she did not give general directions for their use. She had to be quite specific in her directions. With a more competent class she would have discussed a problem, suggested a few references, and given the pupils freedom to study for themselves in class or at home. With this class she supervised very carefully by directing the members to particular references to answer particular questions.

In discussions she often had to answer questions herself. But every few minutes some computation was necessary and this she left to the pupils. The pupils participated in the discussion as they were able but they were all expected to do the computations. The better pupils and the teacher carried the load of the thinking without any attempt to impress the details on the slower pupils. This part of the period took on the aspect of social cooperation.

Thus each pupil was being treated according to his ability. The better pupils got a clearer picture of the whole unit than the poorer pupils. All of them learned that arithmetic is valuable in everyday living though some of them did not nor could not carry through the complex thinking to decide what to do at each step. And all of them did the computations. Two of the important objectives of general mathematics were met—selling the value of arithmetic and perfecting the methods of computation.

Other Non-Academic Courses. Limits of space do not allow us to go into detail with other courses of the so-called second track. Some of the large cities have many such courses, based upon industrial and commercial as well as more personal everyday needs. Smaller systems cannot be so lavish. Many schools now have a general mathematics course in the tenth year. This could well have a core of geometry studied on a lower level of learning than that needed in the traditional academic course. Geometry through experience is valuable to many who cannot get it through logical reasoning. The last year of the senior high school is an excellent place for Consumer Mathematics. In that year the students can realize their deficiencies in arithmetic and can take a keener interest in such things as wise buying, insurance, and taxes.

SUMMARY

The fact of variability in the traits of individuals was known even in ancient times but it has been left to the last half century to deal in detail with the vast problem of individual differences in the schools. Even now we know more about the existence of individual differences than we do about the manner of dealing with them.

In Grades I through VIII, all children should study the same topics in mathematics in order that they may have the same fundamental background. Differentiation should be in levels of learning and in depth and scope. Beginning with the ninth grade, differentiation may be made in courses as well as in these two other ways. General mathematics instead of algebra is better fitted to the needs of a large group of pupils. Those going into industry and commerce as well as those going to institutions of higher learning to specialize in non-mathematical subjects may well take courses more fitted to their needs than the highly technical sequential courses.

Grouping, whether by classes or within classes, is only a means toward an end. The narrower the range of ability within a class, the more efficiently the teacher can take care of the various levels of ability.

10. *Planned Instruction*

Irving Allen Dodes

The purpose of this chapter is to describe and illustrate convenient methods for the planning of instruction in mathematics. It is intended for the practical use of the classroom teacher. For this reason, questions of broad planning, of syllabus change, of the large philosophy of teaching and other such problems will not be touched upon. In the main, the chapter will deal with teaching the content of the daily lesson or of a small unit consisting of a few daily lessons.

Need for Planning. According to modern psychology (field theory), no action takes place without a goal. Hence, there is no such thing as an unplanned lesson. We shall define the "planned lesson" as one in which the goals have been consciously set by the teacher.

When he teaches the class, his goal is not "the binomial theorem" but rather "to review the expansions of $(x + y)^2$ and $(x + y)^3$; to progress by induction to $(x + y)^4$ and $(x + y)^5$; to develop a general rule for expansion of binomials; to apply it to $(a + b)^8$, $(m - p)^6$, and $(2x - 3y)^5$.

That this sort of planning is present in good teaching is an intuitive, if unproved, truth. The teacher needs a plan in order to give direction to the lesson and to make him aware of the relative importance of various items. It is also true that students seem to "sense" an absence of preparation or planning on the part of the teacher. Perhaps they recognize the cues of momentary indecision, or maybe it is an undifferentiated perception on their part. At any rate, this recognition of a lack of conscious goal appears to hamper the student in his search for the meaning of the lesson.

Written Lesson Plans. There are many advantages to a written lesson plan. It is evident that it makes the goals of the lesson definite. Psychologically, this has the effect of "making them approach," i.e., they become easier to reach. In addition, when the teacher has written the lesson plan, he has mentally practiced his procedure. The competent bridge player does the same thing when he bids; he plays the hand mentally. It may not work out as planned, all the time, but in the long run he has a decided advantage over the tyro who knows a formal set of rules but does not foresee the consequences of each action. A further advantage of the written lesson plan is that it permits self-criticism. After a mediocre lesson, the teacher may review his plan and discover wherein he failed to take full advantage of his opportunities. After a good lesson, he may be able to find out why it was so successful. The written lesson plan is sometimes helpful to supervisors. Sometimes the supervisor who watches a lesson is unable to tell what the teacher had in mind; a glance at the lesson plan may aid him in his advice to the teacher.

Although the written lesson plan has all these advantages, it cannot be denied that there are a few disadvantages. A teacher who has spent a half-hour writing a lesson plan may be loath to discard it when the class has developed other plans just as good. It is obviously unwise for the teacher to proceed in one direction while the class is headed elsewhere. Another disadvantage of the lesson plan is that it is essentially static, whereas the teaching and learning process is dynamic. If the teacher regards the lesson plan as a guide (rather than as a crutch), the lesson will be benefited rather than harmed by the written lesson plan. A third disadvantage is that teachers may tend to collect and preserve lesson plans from class to class, and term to term. Since the teacher-pupil-lesson relationship has so many variables, it is rather improbable that the same situation will be reproduced term after term.

Probably the best compromise is the use of a brief written lesson plan after the teacher has gained experience in the planning of instruction. The methods presented in this chapter are given in detail for the beginner but may readily be altered for the use of the experienced teacher.

The Variables in Planning Instruction. There are five main factors in the planning of instruction: the personality of the teacher, the personality of the class, the nature of the lesson to be taught, yesterday's lesson, and tomorrow's lesson. All must be taken into consideration by the teacher who plans his lesson.

First, it is evident that some teachers have a preference for one "type" of lesson as compared to another. Some like the experimental method, some the deductive, some the inductive, some the heuristic lesson; there are teachers who depend mainly upon the blackboard, whereas others intersperse models, films, and filmstrips liberally. One group of teachers emphasizes pupil (physical) activity, whereas others place everything on an intellectual plane. There are, in short, as many preferences and variations as there are teachers, classes, and lessons. *No one method guarantees success.* Instruction which does not fit the teacher is apt to feel "forced." The lesson, even if planned perfectly on paper, would not "flow."

Second, it is obvious that classes differ greatly in their collective personalities. Some classes prefer a lengthy exposure to concrete illustrations before going to the abstract; others are impatient to arrive at the abstraction and deplore a prolonged study of the concrete example. Again, some classes require careful motivating techniques, whereas others are driven by inner fires. Some pupils like to deduce a generalization, then apply it to specific cases; others prefer to use the experimental method to arrive at the generalization. Some classes will ask for more practice problems of the same type; others will become tired of sets of problems or exercises which seem repetitive. Finally, some classes are able to "dig out" the lesson in a supervised study set-up; others are unable to learn in this manner.

Third, the nature of the lesson to be taught is an important factor in the planning of instruction.

A lesson on congruent triangles lends itself well to the experimental or model approach; one on slide rules to demonstration and films; one on volumes to a study of models; one on indirect method to a study of reasoning in so-called non-mathematical situations.

Finally, the lesson for today must continue from the lesson for yesterday and must supply some motivation for the lesson for tomorrow. This may be done in several ways:

Today's lesson may arise from a homework problem, or from a question asked by a pupil on the previous day, or from a report made by a student or a committee of students. It may be a "natural" continuation: if, for example, addition, subtraction, and multiplication have already been discussed in an algebra class, it is natural for the class to expect to do division next. Often, today's lesson will supply the link for tomorrow's. The student who asks, "What would we do if . . . ?" is often asking a logical question for the next day's lesson. The teacher should compliment the student for his perception of the pattern of instruction and, if possible, start with that same question on the next day.

The purpose in providing continuity is to emphasize the fact that mathematics is a complete pattern in which the individual lessons are merely experiences provided to induce insight.

It seems reasonable to conclude that *no fixed set of lesson plans will fit every teacher, every class, and every lesson.* The teacher may, after years of experience, work out a cherished set of plans; but copious alterations will be in order whenever they are to be used.

The "Basic" Lesson. Although there are so many variables in the construction of a lesson, most lessons are probably of the developmental type. Most teachers seem to regard this type of lesson as the fullest expression of their art; and it is very likely true that almost any topic in mathematics may be taught successfully by the developmental method.

In this method, the teacher first motivates the class in order to energize the learning process. Once the pupils have been made aware of the goals, the teacher ties the topic to previous lessons in the subject and to the pupils' experiences. Following this, the teacher, usually by progressive questioning, directs the attention of the class to sub-problems until the main problem has been solved to the satisfaction of the class. When this is done, generalizations and summaries are made, and the class applies its solution to specific problems.

It would seem that this is essentially the Herbartian-step method of teaching. Differences in the actual planning and teaching arise, to be sure, in the application of these steps, rather than in the steps themselves. A developmental lesson taught from the "connectionist" (S—R bond) viewpoint is very different from a lesson taught by the "field" (gestalt) theory.

The main portion of this chapter will be devoted to methods of planning this "basic" lesson from the field theory point of view.

The last part of the chapter is set aside for a discussion of the numerous variations on the basic plan.

PLANNING THE "BASIC" LESSON

It is characteristic of new teachers and teachers-in-training that they possess a great deal of undirected enthusiasm. Their lesson plans are complex, lengthy and usually quite impossible to fulfill. One reason for this is their methods of planning. The beginner starts with a profound anxiety about motivating the lesson. When he has finally created an edifice of motivation, he attempts to stuff the content of the lesson into its attic. (Fortunately, it usually protrudes enough to permit rescue—in the next lesson—by the regular teacher.)

The proper beginning in lesson planning is an analysis of the aims and objectives of the lesson. When the beginning has been planned, the ending should be based upon this, immediately; for it is clear that the problems uncovered at the beginning must be resolved at the end.

Only after the beginning and the end of the lesson have been determined can the teacher efficiently determine the path to be followed from one to the other.

Step I. The Aim. The first step of the lesson plan is, accordingly, a determination of the "aim." For all practical purposes, this does not refer to the larger aims of education or of mathematics, but to the topic given in the course of study, e.g.:

Rate-time-distance problems with one unknown.
Two points, each equidistant from the ends of a line segment.
Introductory lesson on the linear graph.
Sum and product of the roots of a quadratic equation.

Step II. Listing the Objectives. After his initial glance at the "aim," the teacher must differentiate the specific knowledges, skills, and concepts involved.

This is clearly the most difficult part of the plan, especially for the beginning teacher, because it involves not only a recognition of the elements of the lesson, but also an understanding of the barriers involved in the learning process. However, it is the part of the lesson plan which often can be saved from term to term without too much alteration.

Rate-time-distance
  Meaning of rate, time, distance
  Units of these
  Relationship ($RT = D$)
  Distance diagram
  Procedure for solving
  Solution of equation
  Checking solution

Graph of straight line
  Points and number-pairs
  Meaning: axis and origin, etc.
  Plotting points
  Table of values
  Locus concept
  Number of points necessary for line

Two points etc.
  Meaning of equidistance from two points
  Meaning of perpendicular, of bisector
  How to prove segments equal
  How to prove lines perpendicular
  Planning a proof
  The proof
  Application of the theorem

Sum and product of roots
  Meaning of sum, product, roots
  Induction to arrive at hypothesis
  Proof of relationship
  Use in checking answers
  Use in forming equations

Step III. The Terminal Summary. Once the specific objectives have been listed, it is sound psychology to prepare the terminal summary immediately. This has the effect of eliminating impossible goals and of making the path of the lesson specific. In some cases, a terminal summary is provided by the solution of a problem, by a conclusion (such as the statement of a theorem, with explanations), or by a construction. More often, it is in the form of a series of questions based specifically upon the list of objectives:

Motion problems:
1. A certain story has it that a boy scout walked 6 miles in 2 hours. What was his rate? his time? his distance?
2. What is the basic relationship between rate, time, and distance?
3. What precaution must be taken with reference to "units of measurement" in applying this relationship? Give an example.
4. How would you make "distance diagrams" for the following problems?

PLANNED INSTRUCTION

   a. In the explosion of an atom, one particle flies in one direction, another in the opposite direction.
   b. Jack was much faster than Pete. In a race, Pete was allowed to start 30 seconds before Jack, but Jack caught up with him after a while.
   c. When Pam entered the playground, she saw her brother, Lance, at the other end. They ran to meet each other. They met one-third of the distance from the entrance.
   d. The Tigers and the West Branch football teams were lined up at their goal-posts to take pictures when it was decided to change sides. Each team raced to the other goal-post. The Tigers got to theirs first.
5. After drawing the distance diagram, how do you represent the unknown quantities?
6. How is the information in a motion problem tabulated?
7. How do you use this tabulation to form an equation?
8. After solving the problem, how do you determine whether or not the solution is consistent with the conditions given?

Sum and product:
1. What is meant by the phrases, "sum of the roots," "product of the roots"?
2. In the following equations, what is the sum of the roots? the product?

$$x^2 - 3x + 5 = 0$$
$$x^2 - 3x = k$$
$$x^2 - 3x = 0$$
$$2x^2 - 5x + 7 = 0$$
$$3x^2 + 10 = 27x$$

3. How would you check the following approximate answers: 2.29 and -0.54 for the equation $4x^2 - 7x - 5 = 0$?
4. How would you write an equation which has the roots: 1 and 4, 3 and 5/6, 7/3 and 5/3, 4/5 and 6/7?
5. What are two uses for the sum and product relationship?

It is often advisable to terminate the summary with a discussion question which will provide a smooth and pleasant ending for the lesson.

Why do you suppose motion problems are so important in navigation? Why do you think scientists have to know how to "make" equations?

Step IV. Planning Supplementary Aids. When the objectives of the lesson have been listed, and the terminal summary prepared, departmental resources should be explored in order to find appropriate teaching aids. Short films, filmstrips, models, cartoons, editorials, globes, maps, instruments, and tools, are all grist for the mill.

The choice of teaching aid is quite important, for while a good teaching aid does not guarantee a good lesson it is quite possible for a poor one to spoil an otherwise successful learning situation. For example, a mathematics film which has errors will destroy the worth of the lesson no matter what the other good features are.

Supplementary aids in algebra may consist of certain materials related specifically to the topic, such as bankbooks (for interest problems); stock certificates (for business problems); geometric models, balance scales (for the equation); the spring balance (Hooke's Law); and yardstick with sliders (for positive and negative numbers). Supplementary aids in geometry and trigonometry include folding rulers with rubber band attachments, photographs and enlargements (similar figures); a knotted cord (Pythagorean theorem); surveying instruments, and such.

In some cases, the supplementary aids will be found in the departmental closet. In others, the teacher will desire to fashion them, himself; or the student may take part in their manufacture or may bring things from home:

In the lesson on motion problems, toy cars or trains are most effective in pictorializing the situation. In the lesson on "two points equi-distant from the ends of a line segment . . . ," a bow-and-arrow may be used. Tinkertoys are a rich source of material, as are Erector sets.

Step V. Planning the Motivation. In order to explain the method of planning motivation, it will be necessary to make some introductory remarks concerning it.

Motivation may be defined as any stimulus which causes a rise in body energy. For example, hunger, fear, and the injection of glandular extracts are means of motivation. To be sure, they are useless for teaching purposes because they are, in general, *undirected*

Methods like punishment, sarcasm, rivalry, bitter competiton, reproof, and failure are *directed* methods of motivation which operate because they reduce the problem to a level easily understood by the student. (For example, the student may not understand the importance of the goal of the lesson, but he will understand the importance of avoiding the teacher's anger.) These methods are, however, poor for another reason. The real goal, in this case, is avoiding the displeasure of the teacher; the substitute goal brings about insights which are loaded with a pattern of dislike for the teacher and the subject. Long after the lesson has been forgotten (the lesson is the secondary achievement, here), the dislike of teacher and subject (the primary achievement) will remain.

Similarly, praise, reference to marks, parental approval, "races," mathematics tournaments, and the like, serve as motivation because they supply substitute goals which are easier to perceive than the true goal of the lesson. In this case, the secondary achievement (the lesson, itself) is not unpleasantly loaded; but it should be kept in mind that the primary goal was not a mathematical one.

On the other hand, referring to previous lessons, to hobbies, to sports, to the outside interests of children, and to other subjects are samples of good motivation because they utilize the psychological principle of closure, i.e., they tend to fit the new idea into a pre-existing pattern of ideas. The use of models and supplementary aids is also good motivation because it is *directed* and because it reduces the lesson to a level easily understood by the student (it seems to make the goal nearer because perception is better).

Before going to actual examples of motivating techniques, it might be well to say a word about excessive motivation. It seems to be far better to have insufficient motivation than excessive, so far as the teaching and learning situation is concerned. If the lesson is otherwise good, the pupil will become motivated, anyhow. Excessive motivation brings about an effect like over-excitement in small children (the irradiation effect).

For example, a certain teacher motivated a lesson in probability by calculating the odds in an actual dice game played in class. While this was a fairly attractive example of a simple calculation of odds, it was unsatisfactory because the class refused to settle down to the real lesson, afterwards.

In another class, the teacher went to a great deal of trouble to set up a chemical experiment involving a small explosion in order to motivate a lesson in the solution of chemical equation problems. To his utter dismay, he found that the lesson was completely out of hand.

In planning a motivation, the following items should be kept in mind: (a) it should not consume more than a minute or two unless it is an actual part of the lesson; (b) it should refer to the actual goal of the lesson, if this is at all possible; (c) it should make use of the laws of modern psychology.

The three "laws" described below have been postulated in this chapter for the sake of convenience in separating types of motivations. Actually, there is much overlap among these "laws":

Law I: The Law of Natural Closure. When a student has learned a certain amount of mathematics, there is a psychological desire to "fill out" the pattern whenever the incompleteness of his learned pattern is made clear to him.

Law II: The Law of Apparently Near Goals. A student will tend to respond to a challenge when it is not too difficult, even if he is not really interested in the topic. Thus, increasing the understanding of the goal by increasing the perception of the student will tend to "make the goal approach." When it is apparently near, the student will respond. (This is like putting on eye glasses, or using field glasses. The object is not really nearer; it is merely clearer.)

Law III: The Law of Substitute Goals. When a student can be shown that the goal is part of a pattern of other interests or desires, the law of closure begins to operate.

Motivations under Law I. In the following examples, the motivations are clearly an integral part of the lesson. They may, therefore, be called intrinsic. They operate most successfully when the teacher and class are both enthusiastic about the subject.

1. Roots and coefficients: One of the problems assigned for homework was the solution of a quadratic by formula. This problem had been put

PLANNED INSTRUCTION   313

on the black board by the request of the teacher. The teacher asked how the approximate answers could be checked. The class answered, "By substitution in the original." The teacher praised the class, then said, "That would be a good method for checking, but today we're going to learn a method which is easier, faster and just as accurate. Besides, it will work for exact roots and complex roots, just as well."

2. Second lesson on logarithms: "We have already practiced with logarithms to base 2, 3, and 5. Today we will investigate logarithms to the base 10 to see what the advantages and disadvantages are."

3. Parallel lines: "The class has spent a lot of time on lines that cross. Is it possible for a pair of lines on a flat surface to continue forever without ever crossing?"

4. Binomial theorem: The teacher had the aim of teaching the class to solve problems of the type $(ax \pm by)^n$. When the class entered, there was a problem on each front board: $(x + y)^2$, $(x + y)^3$, $(x + y)^4$. The class had been trained to start on "seat problems" as soon as they arrived. The first two were soon put on the board by volunteers, but the teacher interrupted the class while it was struggling with the fourth power. "I don't like to interrupt you, but I can do this one mentally." A student said, "Well, you memorized it!" "All right," said the teacher, "name another number to use as exponent." The boy said, "32." This led into a discussion of the number of terms to be expected if the proposed expansion were done and then into the usual induction. The class agreed to ask for $(x + y)^{12}$ which the teacher did.

5. Solving the quadratic by formula: "One of the important functions of mathematics is to solve equations. You have already solved equations like $x - 4 = 0$ and $2x - 3 = 3x + 5$. Equations of this type are called linear equations for reasons which we discussed previously. You have also learned how to solve certain quadratics like $x^2 - 6x + 9 = 0$. How was that solved? Can you think of a situation where you could not use this method to solve a quadratic?"

6. First lesson on circles: "You have already studied figures formed by two lines. What do you call such figures? (angles, pair of parallels). You have also studied closed figures formed by three lines. What are they called? Today, we're going to begin the study of a very important closed figure formed by only one line. Can you guess what it is?"

Motivations under Law II. When the objectives of the lesson are so difficult or involved that the students may have difficulty in understanding the problem, the motivation should be of a type which makes the problem clear:

1. Overtake problems: Two boys were asked to go to the front of the room. They were placed side by side at one end. One boy was asked to start walking across the front of the room. After he had walked five or six steps, the other boy was asked to overtake him.

2. Overtake problems: "Last year, after a big steamship left the pier, it was discovered that a very important passenger had been left behind. Fortunately, the ship was not very far away, so he hired the owner of a speedboat to take him to the ship."

3. Locus: The teacher tied a piece of chalk securely to a piece of string and swung the chalk rapidly. "What geometric figure does this look like?"

4. The sine curve: The teacher showed a film called "Periodic Motion."

5. Congruent triangles: The teacher had cut a triangle out of cardboard and asked what measurements would have to be made to make another triangle exactly like the one he had.

6. Similar triangles: The teacher brought in a 4 x 5 and an 8 x 10 photograph of his son. The teacher and class discussed reasons why the two photos were similar.

Most lessons which employ supplementary aids are using this law of motivation. "Concretizing" the problem tends to make perception easier; thus uses Law II.

Motivations under Law III. The third law must be used in classes where considerable energization is necessary. Here the true goal appears too "distant," so a substitute goal is provided. The substitute goal may be based upon current events, comic strips, puzzles, games, recreations, competitions, and tests. Or a novel situation may serve as motivation: a student may, for example, take over the class (this is like a game). Committees may be set up; here the group activity serves as motivation. Physical activity, such as board work, making models, may serve to energize the instruction. A few examples are given in the following:

1. Motion problems: "How many of you have seen jet planes? At what rate does a jet plane travel? You can understand why some people call this the Age of Speed."

2. Two points each equidistant: On the previous day, the "extra credit" assignment was made to read about the crossbow and to report

PLANNED INSTRUCTION   315

to the class. The report took two minutes. The teacher asked, "How must the arrow be placed in order to be sure it flies straight?"

3. Interest problem: The teacher wrote the following problem on the board before the class entered: "When this school was built in 1915, a student by the name of Smith deposited $1 in the school bank. If the interest rate is 3% compounded quarterly, how much would you pay for Smith's bank account? Guess!" The estimates ran as high as $10.

4. Graphs: "How many of you have been reading about the new experiments with small atom bombs? A great mathematician now at Princeton worked on the basic Relativity Theory which made the atom bomb possible. Does anyone here know his name? As you probably know, Einstein's work had to do with space. He showed that you can use numbers to tell about things in space. Today we're going to do the same thing on a small scale—we're going to learn how to tell the position of things by means of numbers."

5. Binomial theorem: The teacher had as his aim the illustration of the power of the inductive method of discovering mathematical theorems. "How many of you read science-fiction novels? Recently I read a story in which a scientist was asked to find the volume of a four-dimensional figure with equal sides and angles. At first he didn't see how it could be done, but after a while he began to think about the formulas for the area of a square and the volume of a cube. Then it came to him in a flash. Can you guess how he figured it out?" After a student "got it," the teacher continued, "You have just used a method of reasoning called "induction." We're going to apply the same method to find an easy way to solve problems like $(a + b)^6$ or $(a - b)^9$. Can you think how to start our induction?"

Step VI: Planning the Development. You have now (a) determined the aim of the lesson, (b) listed your objectives, (c) composed a terminal summary to fit the objectives listed, (d) surveyed the supplementary aids which might be appropriate in your lesson, and (e) planned your motivation. It now becomes necessary to decide how to channel the energy of the class to resolve the problems which have been made clear and desirable to it.

It is during the development of the lesson that learning takes place. According to field theory, after a supply of energy has been made available, and after attention is directed to the goal (or substitute goal), the phenomenon known as "perception" occurs.

The student now perceives at least one aspect of the goal. After a sufficient interval of perception, the phenomenon known as "insight" takes place. Here the student recognizes the pattern which constitutes the goal. This insight is the aim of the development.

Since insight follows perception, it is necessary to ensure efficient perception in order to ensure learning. This may be done in several ways:

1. Perception may be sharpened by high motivation.
2. The time allotted to perception may be increased by going slowly and by giving sufficient time for thought.
3. The abstraction which it is desired to teach may be introduced by a concrete illustration which is more easily perceived.
4. Where the "whole" goal is too large to be perceived by the student, smaller "wholes" may be taught, then integrated. This will be called the method of "progressive wholes."

In the usual type of lesson, development is accomplished by the question-and-answer method. If the atmosphere of the class is a good one, there will ordinarily be enough cross-talk and general discussion to cause the formation of insightful relationships. The teacher should plan a few key questions which will keep the class on the desired path.

Of course, questions which have perfect form do not guarantee a good development; and questions which are considered to be in poor form do not necessarily cause any damage. However, questions of the following types are usually frowned upon:

> These two lines are ... (incomplete question)
> What do you think about this triangle? (vague question)
> Is the sum $3x$? (yes-no answer)
> When you add 2, the result is what? (surprise ending)
> How many think this is right? (the vote question)
> How do you construct this line and what does it intersect? (double question)

Good questioning refers to questions which offer an amount of challenge which is just right. Questions which are too easy are sometimes met by disdainful silence and at other times by chorus answers. Questions which are too hard cause discouragement on the part of teacher and class. Between these two extremes lies the pattern of good questioning. Experience has shown that a liberal interspersing of "why" and "how" questions in the development serves to stimulate discussion and thought. In the following examples the "why" and "how" questions have been omitted, because it would be unnecessary to write them in a lesson plan. It is understood that they are to be asked.

1. Motion problems: At what rate does an ordinary airplane travel? What distance would it travel in 30 minutes? Suppose a car is traveling at 30 miles per hour; what distance would it travel in 4 hours? in 2 hours? in x hours? Suppose two trains start at the same place at the same time. One goes at the rate of 40 mph due north, the other at 50 mph due south. How far apart are they after one hour? after 2 hours? after x hours?

2. Exponential equations by logarithms: The class had solved $3^x = 3^7$ and $2^y = 2^3$. They were given $2^y = 8$ to solve. This was a motivation using Law I. The teacher asked: "How would you solve $2^y = 8$? Can you think of an exponential equation something like this which could not be done so easily by this method? Try to estimate the approximate value of x in the equation $2^x = 7$. This means that the answer is 2-point-something. What does this sort of exponent (pointing to 2.5 which was the guess) remind you of?" This was technically a vague question, but every hand went up and the class realized that logarithms were involved.

3. First lesson on circles: "If you wished to construct another circle equal to this one, what would you measure? Consider this circle with the 5-inch radius. With respect to the circumference, where would a point three inches from the center be? 8 inches? 5 inches? What are your conclusions with respect to distance from the center and the circumference? Here are two equal circles, O and O'. Mark off equal arcs AB and $A'B'$ and draw the radii. What would you expect to be the relationship between angles AOB and $A'O'B'$? What is one method of proving two angles equal in two equal circles? What do you think is true about chords AB and $A'B'$ in these circles? How do you usually prove two line segments equal? But there are no triangles here! How would you draw lines to make the triangles which you mentioned."

In the above, the developmental questions have been made far more detailed than would be necessary or desirable on a written lesson plan. Usually a cue or brief note is sufficient to indicate the course of the development:

4. Discriminants: The class had been separated into four parts and each part was assigned to apply the quadratic formula to one of the following:

$$x^2 + 6x + 7 = 0$$
$$x^2 + 6x + 8 = 0$$
$$x^2 + 6x + 9 = 0$$
$$x^2 + 6x + 10 = 0$$

They were asked to leave the answer in radical form. The radicals were $\sqrt{8}$, $\sqrt{4}$, $\sqrt{0}$, $\sqrt{-4}$. The cues on the lesson plan were:
  a. Solutions on board
  b. Teacher and class solve all except last
  c. Speculation about $\sqrt{-4}$
  d. Imaginary axis, "i"
  e. Why the discriminant shows type of solution.

There has been much speculation regarding the relative advantages of (a) logical vs. psychological, (b) inductive vs. deductive, (c) why vs. how, and other schemes of development. This has led to experimentation with the following conclusions:
  1. It is quite true that some teachers are consistently successful whereas others are not,
  2. Yet repeated experimentation has failed to show any advantage of one type of development over any other, and
  3. Expert analysis has failed to show that any specific teaching act or any specific teaching method is significantly correlated with this success or lack of success.

Under the circumstances, the teacher can only draw upon the generalizations of educational psychology in planning his development, in the hope that these are more likely to lead to success than a haphazard approach. The crux of the matter, in planning the development, would seem to be the nature of the class. In a bright, interested class, it is perfectly possible to pursue a strictly logical sequence:

What theorem have you proved concerning the angles opposite the equal sides of an isosceles triangle? What does this suggest for a triangle which is not isosceles? What theorem can be used to prove that one angle is larger than another? What line can be drawn in this triangle to make angle C a "remote interior angle"? (There are a number of such lines, of course; the class is now led, along logical lines, to choose the useful one.)

In a class which is unable or unwilling to integrate the lesson into its own pattern of pre-existing interests or concepts, the so-called psychological development must be used. This psychological development is, usually, another logical development in which the steps are shorter. In some cases, steps are rearranged. In others, the difficult steps are postulated or are covered by easy analogies.

What theorem have we proved concerning two angles of an isosceles triangle? Now look at this scalene triangle which I have cut out of paper. Are the angles equal? Which is the largest angle? the smallest angle? (Measurement with protractor may be done.) Which is the largest side? the smallest side? What general conclusion do you think can be drawn? Let's fold the paper so that AB falls along AC, like this. Why doesn't AB stick out? Now look at angle ADT and angle C. Which is larger? How do you know? Suppose you have started with this folded figure. How could you tell that angle ADT must be larger than angle C in the small triangle?

The psychological development takes a great deal more time, as well as trouble, because there are so many short steps. In writing the plan of such a lengthy development, the teacher might note just a few key phrases:

Isosceles triangle comparison
Paper scalene triangle (get paper for class)
Fold paper
Elicit construction lines and proof

He might also note two or three questions which he thought would direct the class properly.

Similar considerations hold with reference to inductive vs. deductive developments. A bright class will take in its stride the proof that the sum of the angles of a triangle is 180 degrees. In another class, an experimental or statistical study will lead to the concept desired. It will still remain necessary to prove the theorem, but the goal will be understood more clearly. In some cases, it is advantageous to use the inductive technique even for bright classes: For example, even a bright class has trouble learning about exponents, deductively. How much easier it is to use induction!

Simplify $5 \cdot 5 \cdot 5 \cdot 5 \cdot 5/5 \cdot 5$ (Then a few more examples); How would you simplify $x \cdot x \cdot x \cdot x \cdot x/x \cdot x \cdot x \cdot x$? Now simplify x^6/x^4? What general rule can you make . . . ? What is x^6/y^2? In the light of this example, how must you modify the general rule which we wrote on the board? Apply your rule to $2^6/2^4$; $2^6/2^5$; $2^6/2^6$; $2^6/2^7$; $2^6/2^8$. But 2^6 is 64 and $2^6/2^6$ actually means 64/64. How much is that? What is your conclusion about 2^0?

It is rather evident that the deductive, logical method is the easier and more direct development wherever the nature of the lesson warrants its use. Whenever the goal is so complex, however, that the student cannot perceive its true pattern all at once, it is probably advisable to plan a psychological and/or inductive development. In this case, learning still takes place by "wholes," but the method of "progressive wholes" is used. In this, smaller "wholes" are learned, according to the capacity of the student, and then these are integrated as desired.

Very much the same considerations apply to the why-how controversy. If the class is able to perceive "why" before "how," this is clearly the most satisfactory scheme of development. However, many topics (such as "square root by the algorism") are so difficult to understand that there is little or no possibility of deducing a method before applying it. If "why" precedes "how," it means that the method must be developed by deduction. In the case of a topic as difficult as this, there is no help for it. The "how" is shown, and then justified immediately by multiplying the square root by itself to obtain the original number. When the "why" follows the "how," it may consist of a "justification" rather than a "proof."

Step VII: Planning the Board Work. Whereas the notebook is a student's own summary of important points, the blackboard is the notebook for the entire class. On it is a running summary of the lesson, so that the class always has before it the goal and the steps leading to the goal.

In planning a lesson, the teacher must take into account the physical limitations of the class blackboard. He must think of the placement of the various items, consider what is to be kept and what is to be erased. In some cases, teachers may make a rough sketch or note of the plan of utilization of blackboard space. Such a plan is shown at the end of this section.

Step VIII: Planning Drill and Generalizations. The lesson plan, written or mentally planned, is now almost complete. From the viewpoint of the student, he now knows what the purpose of the lesson was and he has gained some insight into the knowledge, concept, or skill which was the content of the lesson.

The psychological steps which follow are (a) differentiation, (b) integration, and (c) the attainment of precision. These take place as a result of drill and generalization when these are properly planned and accomplished.

"Differentiation" refers to the fact that learning takes place by "wholes." You learn a person's face before you learn the color of his eyes. Students learn the theorem about inscribed angles before they discover that an angle inscribed in a semicircle is a right angle. The teacher hurries the process of differentiation by directing attention to corollaries, applications, and exercises. The lesson plan should, therefore, include problems which cause the small details of the large pattern to emerge. These may be graded problems of the usual type, theorems and corollaries, or applications of the concept to a different situation:

1. After teaching the solution of the exponential equation by logarithms, the teacher asks the class to differentiate between $\log (8/2)$ and $(\log 8)/(\log 2)$.
2. After teaching the binomial theorem, the teacher asks the class to use it to find $(1.02)^6$.
3. After teaching infinite geometric progressions, the teacher applies it (a) to an analogous geometrical problem, and (b) to repeating decimals.
4. After a lesson on areas in plane geometry, the class is given a figure made up of different geometrical figures.
5. After teaching the indirect method of proof, the teacher calls attention to life situations requiring this method, such as circumstantial evidence.
6. After a lesson on numerical trigonometry, the teacher has the class measure the height of the school flagpole using their protractors and a yardstick.

"Integration" refers to the fact that learned material gradually fits itself into other patterns. For example, at first the inscribed angle theorem stands by itself as an isolated "whole," but under the skillful direction of the teacher, the unity of pattern for chord-chord angles, inscribed angles, and secant-secant angles becomes apparent.

Almost always, the same drill which causes differentiation brings about integration, the reason being that the student sees the pattern from a different point of view. The teacher may assist the process of generalization by providing appropriate experiences:

1. After studying the straight line graph, the class is led to think about points off the line (locus idea).

2. After teaching the chord-chord angle, the teacher treats the central angle formed by two diameters as a chord-chord angle.

3. After teaching the quadratic formula, the teacher brings about a situation in which a quadratic done previously by factoring is now done by the formula.

4. After teaching the "digit" problem, the teacher discusses the binary system.

5. After teaching complex numbers, the teacher refers to the three dimensional graphs of second and third degree loci (see Fehr, *Secondary Mathematics*, p. 285-91).

6. After teaching plane loci, the teacher allows the class to discuss solid loci briefly.

The third function of drill is to bring about "precision." Often this is the only thing consciously aimed at. However, since it is a normal consequent of differentiation and integration, there is no great need to discuss it separately.

Two important things to remember in planning drill are:

1. Drill must be motivated. If the lesson was well done, the need for drill will be apparent to the class and no further motivation will be necessary. If the lesson was not sufficiently successful, motivation can be brought about by Law II; i.e., by grading the drill work in such a way that only a slight challenge is offered. If the lesson was poorly received, Law III must be invoked; e.g., "Problems like this will be on your next test!" The method of motivation of drill is, in reality, a key to the opinion a teacher has of his own lesson.

2. Drill must provide for differentiation, integration, and precision. The method for accomplishing these three objectives of drill has been explained above.

Step IX: Planning the Assignment. Most teachers regard the homework assignment as an important part of the mathematics lesson. Experimental evidence on this question has been far from convincing, although it appears to favor somewhat the homework groups. However, as a practical matter, teachers are expected to assign homework. The question arises how this should be planned.

Three methods of assigning homework are (a) the repetitive type, (b) the voluntary type, and (c) the spiral type.

The repetitive type of homework assignment is basically contrary to the tenets of field theory, although it is in full accord with stimulus-response psychologies. According to modern psychology, learning takes place at the first actual perception followed by insight. Repetition of the stimulus has the effect of lengthening the interval of perception; it does not, per se, bring about learning.

The voluntary type of homework assignment, in which the pupil does whatever he thinks is necessary, may possibly be successful where the motivation is high and the pupil knows what he needs. It hardly seems reasonable, however, to expect pupils to be able to recognize their own needs, especially if they do not understand the work. Besides, as a practical matter, it is difficult for the teacher to check or go over voluntary assignments in a class.

The spiral type of homework assignment, developed by Simon L. Berman, chairman of the Mathematics Department at Stuyvesant High School, New York City, is in full accord with the principles of field theory. In this method, the main part of the daily assignment deals with previously learned material, while a small part is devoted to the material just learned. The spiral assignment takes into account the fact that forgetting is a normal part of the phenomenon of learning. After the pattern has been learned; i.e., it has been differentiated and integrated, and a certain amount of precision has been gained, it begins to fade. First, precision is lost. Thus, a student who at one time was very facile in the solution of "originals" in geometry becomes hesitant. After a further interval, the effect of differentiation and integration disappears. The student not only cannot solve an "original" quickly, but he is even unable to reconstruct the pattern of analysis which he once knew. Finally, only the bare bones of the pattern of proof are left. The adult, for example, remembers that there is such a thing as proof, but he is unable to remember the steps.

In the spiral method, the various knowledges, skills and concepts are interwoven into successive assignments in such a way that precision is not lost. Furthermore, problems are brought before the students at intervals in such a way that maturation and growth are encouraged.

For example, suppose Topic A is taught on day one. Problems are assigned from this topic. On the second day, another problem from topic A is assigned. Now one day is skipped. On the fourth day, another problem from topic A is assigned. Now two days are skipped. On the seventh day, another is assigned. Now three days are skipped. On the eleventh day, another is assigned. Now four days are skipped. On the sixteenth day, another is assigned.

This interval is lengthened until it reaches 5 to 10 days, depending on the importance and difficulty of the topic. Then the topic is re-assigned once every week or two.

For the first 20 days, the pattern of assignments would look something like the following before the deletions which will be explained later:

DAY	TOPICS	DAY	TOPICS
1	A	11	AEHJK
2	AB	12	BFIKL
3	BC	13	CGJLM
4	ACD	14	DHKMN
5	BDE	15	EILNO
6	CEF	16	AFOP
7	ADFG	17	BGKNPQ
8	BEGH	18	CHLOQR
9	CFHI	19	DIMPRS
10	DGIJ	20	EJNQST

PLANNED INSTRUCTION 325

In actual practice, the teacher will be influenced by three factors: (a) If topic A leads to topic G, and if the study of topic G involves a review of topic A, then the assignment of topic A may be terminated as soon as topic G is taken up. (b) It will be noted that on certain days (which could, of course, be calculated), topics tend to pile up. In this case, the teacher may either omit one or more of the less important items, or may place the item on a near day where there is no such accumulation. In actual practice, this seldom occurs because of the first factor. (c) A specific assignment may accumulate too many difficult problems, by pure chance. A few interchanges between adjacent days will always solve this problem.

This method, which at first glance seems so complex, is in reality very easy to apply. The following step-by-step procedure shows how the first 15 assignments were made in Plane Geometry II, using the S—S—S textbook.

First step: Rule a notebook into "cells," allowing one cell for each homework assignment. It is convenient to use a hard covered 8" x 10" notebook. Start on a left-hand page and divide each page into three vertical columns and five horizontal columns. Number the cells consecutively from 1 to 60 for a semester course.

Second step: The first topic was "ratio and proportion." This is covered in pages 205–208, with exercises on pages 207 and 208. On the first day, 207/2, 4, 6, and 208/2, 4, 8 were assigned. Then, immediately, 207/7 and 208/5, 10 were written in the second cell, 207/8 and 208/11 in the fourth cell, and so on up to the eleventh cell, in which 208/13 was assigned. No more were assigned from these pages because it was felt that the work on similar triangles would cause review of the concepts of ratio and proportion.

Third step: On each day, when the homework for the new topic was assigned, the main part of the assignment had already been written by the spiral method. The teacher glanced at the problems to be sure the assignment as a whole was not too heavy. After rearrangements, the first 15 assignments were as follows (the work preceding page 207 had been left over from Plane Geometry I from a spiral assignment):

DAY	TOPIC	NEW WORK	BALANCE OF ASSIGNMENT
1	Ratio and prop.	207/2, 4, 8; 208/2, 4, 8	35/1, 2; 41/2
2	Proportionals	209/1; 210/2; 211/2	39/1; 207/7; 208/5, 10
3	Line par. 1 side	213/4, 5	39/2, 3; 207/8
4	Prove lines	218/16, 17	35/4; 41/3; 208/11; 213/6
5	Construction	Construction	41/4; 213/7; 218/3, 4; 215/2a; 216/1
6	$aaa = aaa$	221/1, 4	35/5; 39/5; 208/12
7	Corollaries	222/5, 6; 223/6	41/5; 43/2; 213/10
8	Products	224/2	35/6; 43/3; 218/19; 222/7; 223/7
9	Other methods	227/5	41/6; 51/4; 208/13; 222/6
10	Int. chords	231/2	43/4; 51/2; 222/8; 224/6; 227/6
11	Tan. and sec.	231/3; 232/2; 233/2	43/6; 56/7; 222/7
12	Alt. on hy.	237/3, 4	59/4; 231/4; 232/3
13	Construction	238/3	56/9; 224/7; 233/3; 237/5
14	Pyth. th.	240/4	64/19; 68/2; 232/4; 237/6
15	Special triangles[1]		68/4; 224/8; 223/4; 240/9

[1] Problems from a mimeographed sheet were used for the special triangles.

Of course, it is to be understood that the listing displayed above is an explanatory one; the actual assignment notebook consisted of a series of 60 cells in which the numbers of the problems were entered.

In addition to the problems as shown, there was an optional assignment on each day to intrigue and challenge the brighter student.

The Complete Lesson Plan. The method of planning a basic lesson may be summarized in the following steps:

1. Read the aim and decide what the central purpose of the lesson is.

2. List the objectives carefully. Accomplishment of these objectives must mean accomplishment of the aim of the lesson.

3. (Leave the rest of the first page blank.) On top of the second sheet, prepare the terminal summary which will ensure that the class has realized the objectives of the lesson. The terminal summary must refer specifically to the objectives listed.

4. Go back to the first page and decide on the supplementary aids to be used. The lesson will probably revolve about these, if you decide to use them.

5. Now plan your motivation. Unless the motivation is an integral part of the lesson, plan to consume no more than a minute or two.

6. The development of the lesson should be continuous with the motivation. Match the development to the lesson and the class. A few cue questions and notes are sufficient, in general.

7. It may be helpful to plan the board work briefly.

8. Plan your drill in such a way that it brings about differentiation, integration and precision.

9. Plan your assignment, using the spiral method.

The following condensed record of a lesson is offered to illustrate the result of a lesson plan. The actual reaction of the class was included to make it more readable. Although no claim is made as to the worth of this lesson plan in any other class, it may be stated that the lesson was eminently successful for the teacher and class involved. (No assignment was made in this class because it was a demonstration lesson.)

Record of a Lesson on Graphs. This was a normal class in the first semester of an average high school in Queens, New York.

Motivation: How many of you have been reading about experiments with small atom bombs? (Almost the entire class raised hands. It had been in the morning paper.) A great mathematician now at Princeton worked on the basic Relativity Theory which made the atom bomb possible. Does anyone know his name? [Einstein.]

Development and Drill: As you probably know, Einstein's work had to do with *space*. He showed that you can use numbers to tell about things in space. Today we're going to do the same thing on a small scale—we're going to learn how to tell the position of things by means of numbers.

Suppose we call this row (the middle row), row 0; this one, row 1; and this row, row 2. Write in your notebooks, $R = 0$, $R = 1$, $R = 2$. Now let's see. This is row 2, this is row 1 and this is row 0. What shall we call this row? [−1.] Good. What shall we call this row? [−2.] Good. Be sure to write your row number in your notebook.

Now let's see whether we can identify a person by this number. Will the person with row number 1 please stand? (The entire row stood.) (The teacher appeared chagrined.) Something's wrong! I meant only this girl. What other information would be needed to identify this particular girl? [The seat number.] How many numbers are required to locate a single person in this class? [Two.]

All right, let's number the seats, too, if we really must. This middle line is line 0. This one is line 1. Can anyone help number the other lines of seats? [2, 1, 0, −1, −2.]

Now, let's see whether we can identify a particular person. How many numbers will we need? [Two.] All right, I'll call two numbers; the first one will be your row number, the second your seat number. Please stand for an instant when I call your co-ordinates (that means your two numbers). (The teacher practiced calling on pupils by coordinates.)

How many numbers or co-ordinates are needed to locate a single person? How many numbers does each person have? (On Board I, the teacher wrote:

EACH PERSON IS A NUMBER-PAIR
EACH NUMBER-PAIR IS ONE PERSON

Let's see whether we can locate a point on the board in the same way. This paper will help you do the same thing at your seats. (Handed out graph paper.) It is called graph paper, or quadrille paper, or cross-ruled paper.

Let's draw a line down the middle of the paper and another across the middle. (Teacher demonstrated on graph board.) These two lines are called axes. The Y-axis is the vertical one (teacher showed) and the X-axis is the horizontal one (teacher showed). These axes are marked off in units, like this. Most people who work with these a great deal, like scientists, engineers, draftsmen and navigators, make every fifth line a bit longer, like this. This point, the zero-zero point, is called the origin.

How many numbers do you think will be needed to find a single point on this graph board? [Two.] Very good. Suppose we agree that the first number I call is the x-value or x-coordinate, and the second is the y-coordinate. Just remember that we are calling them in alphabetical order:

first x, then y. What does (3, 2) mean? [$x = 3$, $y = 2$.] Will someone come to the board and show the class where this point is? (The teacher drilled using points in all quadrants.)

How many numbers are required to find a point? How many numbers does each point have? (The teacher went back to Board I and altered it: see the diagram.)

Finding the position of an object is an important use of this method, but there is another use even more important. We can actually show relationships between numbers! Let's go back to your row-and-seat co-ordinates for a moment.

If your first co-ordinate, R, equals your second co-ordinate, S, please stand and remain standing. (On Board II, the teacher wrote $R = S$.) (One pupil made a mistake; the class noticed this immediately. The error was used to emphasize the fact that a straight line was formed.) What does this figure look like? [A straight line.] Now let's go back to our graph chart. Which axis is like the row-axis? [The X-axis, or the one across?] Which axis is the seats-axis? [The Y-axis] (On Board II, under $R = S$, the teacher wrote $x = y$.) Instead of $R = S$, suppose you were asked to draw the "picture" of $x = y$ on the blackboard. What do you think the figure would look like? [A straight line] Will someone come to the blackboard and hold this yardstick across the graph blackboard where he thinks the line would be? (A pupil did it incorrectly.) Well, that's the right idea! Let's check. Will you choose a pair of equal numbers? You? You? (Meanwhile, on Board II, the teacher made a table of values.) Turn your graph paper over and prepare it for a graph by drawing the axes and showing the units. (The teacher did the same thing on Board II.) You see, in this table of values, pairs of numbers. When you think of a number-pair, what comes to mind? [A point.] Let's locate the points which we have already described. (One pupil did it at the board while the others did it at their seats.) Do you know now how the straight line should be drawn? (The class raised hands frantically.) Let's do it. (One pupil was sent to the board to do it.)

How many number-pairs did we use to draw this line? [Four.] Suppose we had only three number-pairs. (The teacher erased one point.) Would we still be able to draw the same line? Suppose we took only two number-pairs? one number-pair? (A pupil said that you needed two number-pairs because otherwise there would be no place to place the other end of the ruler.) How many points, then, do we need to draw the line? [Two.] How many points did we have before? [Four.] Would it be possible to have more points on this line? [Yes.] Will you suggest another point that must be on this line? [(5, 5); (6, 6).] Are there any equal fractions that could be on this line? [(1/2, 1/2), (1/3, 1/3).] Are there any mixed numbers that could be on the line? (The teacher explained what mixed numbers were, and the class supplied many points.)

How many points are there actually on this line? [Indefinite, a lot, "infinite."] Suppose we choose any point on this line. Will you come to the board and put your finger anywhere on the line? What are the coordinates of this point? [It's about $(4\frac{1}{2}, 4\frac{1}{2})$.] What is true about those two numbers? [They're equal.] What is true about every number-pair on this line? [The numbers are equal.]

Now here's a hard one. Suppose we choose a point not on the line, like this one, (4,2). Are the numbers equal? What can you say about any point not on the line? [The numbers are not equal.] Suppose I mention a number-pair like (3,5). Can you tell the class, without looking at the line, whether this point will be on the line? [It won't be on the line because the numbers aren't equal.]

You did so well with that, let's try another relationship. We can use the same set of axes. I'm thinking of a number that's double another number. Can you think of a pair of numbers like that? (On Board II the teacher made a table of values as the number-pairs were offered.) When you see pairs of numbers like these, what do you think of? [Points.] These points happen to lie on a straight line. How many of them do we need to actually draw the straight line? [Two.] True. To be sure, a good mathematician usually takes three points. The third point is taken to check the first two. On our table of values there are seven pairs of numbers. How many points are represented? [Seven.] Choose any three and plot the straight line on your own paper. (The teacher went through the room and finally chose a student to plot the locus on the blackboard.)

Suppose we chose any other point on this line. What would you expect the relationship would be between the two numbers? Suppose a point is taken which is not on the straight line, what would you expect the relationship to be between the two numbers? Suppose a pair like $(-3,7)$ is chosen. Without trying—Is this number pair on the straight line?

These straight-line graphs are called linear graphs, and the relationships that give you linear graphs are called linear equations. One example of a linear equation was $y = x$. What was the other? [$y = 2x$.] Not all equations are linear. At another time you will study graphs which are not straight lines and also straight-line graphs which do not go through the origin like these two.

Summary: Let's put together what we have learned today. How do we represent a pair of numbers on the graph? How many numbers are attached to each point on the graph? How do we represent a linear equation like $y = x$? What do you call the two straight lines that are drawn before the graph is drawn? [Axes.] How many pairs of numbers do we need to have to draw a straight-line graph? What is the third pair of numbers used for? Suppose we took any pair of numbers whatsoever. Would the point necessarily be on this straight line? When would it be on the line? When would it be off the line? Suppose I chose any point on the line $y = 2x$. What would be the relationship between the two numbers? Suppose I chose a point not on $y = 2x$, what could you say about the relationship between the two numbers?

Why do you suppose that the straight line $y = 2x$ is often called a "picture" of the equation $y = 2x$?

Plan for Board Work:

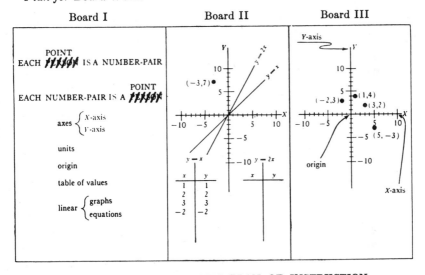

VARIATIONS ON THE BASIC PLAN OF INSTRUCTION

Special-Purpose Lessons. Although education textbooks commonly differentiate between lessons with words and phrases like "supervised study," "laboratory," "lecture," "heuristic," "activity," "recitation," and the like, there is probably no important difference between any two of them so far as learning, itself, is concerned. The learning of mathematics is, after all, a purely mental phenomenon. The physical activity which takes place, whether it involves teacher activity, pupil activity, or the activity of mechanical devices, is of no importance except as it contributes to mental activity.

It would seem to be much more realistic to differentiate lessons in accordance with their main purposes. For example, the first lesson in a certain topic may be an introduction designed to motivate the series of lessons, rather than to convey information:

1. In order to introduce indirect measurement in an elementary algebra class, the teacher took the class to the baseball diamond where the period was spent planning a scale drawing of it. This led to a study of similar figures.

2. In order to introduce areas, one teacher spent the entire period in paper-cutting exercises. Having taken the area of a square by postulate, the class was able to develop, empirically, the formulas for triangles, parallelograms, trapezoids, and regular polygons. The class was prepared for the proofs to be attempted in the next two weeks.

3. In order to introduce logarithms, the teacher had committees report on computing machines. This emphasized the need for efficient methods of performing mathematical operations.

The special purpose of a lesson may be to amass data. An experimental or laboratory type of lesson seldom requires any special motivation since the novelty (in mathematics) of this approach excites the curiosity of the student. The result of the lesson is usually a collection of data which may have a more or less scientific use.

As an introduction to the use of trigonometric tables, the class constructs, using protractors, angles at intervals of five degrees. The sides of the triangles formed are measured with scales, and the sines, cosines and tangents recorded in a "class table of trigonometric functions."

A special lesson may be devoted to drill. This is usually called a "review" lesson, although there is no distinction between "review" (as the word is usually used) and "drill," as it has been illustrated in this chapter.

1. In preparation for an examination, the teacher distributes copies of previous examinations. The class works on these.

2. After a study of trigonometry in the elementary algebra class, the teacher takes the class out to survey a lot. The application of their knowledge constitutes a review.

PLANNED INSTRUCTION

In some cases, the class may spend an entire period in catching up on homework.

It is readily seen that these special-purpose lessons are, in reality, similar to the basic lesson. In these lessons, however, some features have been repressed or eliminated, while others have been emphasized.

Groups of Lessons. There is something to be said for a fixed routine in teaching. A pupil who leaves Mr. X at the end of Plane Geometry I and goes to Mr. Y is often heard to mutter that he was just getting used to Mr. X's method. Routine lends itself to placid class management and thus fosters class discipline.

However, placidity is not an aim of mathematics instruction. Learning takes place because of the disturbance of the pupil's pattern. The teacher can increase this disturbance by including in his repertory of motivations the variation of lesson plans.

One easy way to do this is to plan lessons in groups. The first lesson may, as a whole, correspond to the first part of a basic lesson; the last lesson may, as a whole, correspond to the terminal summary; and there may be one or more intermediate lessons corresponding to development and drill:

A teacher planned a group of five lessons on the slide rule. The first lesson consisted of committee reports on the history of the slide rule and an exhibition of various types of slide rules. The second lesson was a showing of films on the slide rule, followed by a discussion of the films. The third lesson was one in which the teacher demonstrated on a large model while the class operated their own slide rules. The fourth lesson was drill on multiplication and division. The fifth lesson was a test lasting 20 minutes, followed by a review of the test.

Most of the special-purpose lessons are parts of a group. For example, film lessons are always preceded by an introductory lesson and followed by a drill and summary lesson. Supervised study lessons are motivated in some manner, then followed by recitation or test (summary). Laboratory lessons are introduced by another lesson showing the need and purpose of the experimental work, and followed by one or more lessons exploiting the data amassed.

The experienced teacher will eventually come to regard the

entire semester's lessons as a group. This is obviously the psychological operation of integration acting upon the teacher. With this in mind, it is easy to see why the opening lessons in any course are heavy with motivation, whereas the closing lessons are of the review and terminal summary type.

CONCLUSION

In this chapter, the theory and practice of planning instruction has been described and illustrated. The method has been shown in detail for explanatory purposes. Individual users of the method will readily see how to shorten it for their own purposes. The plan may be written (and should be, for new teachers) or may, in many cases, be carried mentally. The method of planning may be applied to individual lessons, to groups of lessons, or even to the lessons for an entire course. The psychological justification for each step was given in terms of modern psychology.

11. Learning Theory and the Improvement of Instruction—A Balanced Program

JOHN R. CLARK AND HOWARD F. FEHR

WAYS TO IMPROVE INSTRUCTION IN MATHEMATICS

No PHYSICIAN is worthy of his profession who is not continuously improving his knowledge of prevention and cure of diseases. As new knowledge and procedures are proved, the physician absorbs them into his practice, changing his techniques and administering new and more potent drugs, to the benefit of his patients. And so it is with law, scientific research, engineering, and all other service professions. In similar manner in the profession of teaching, the alert minded teacher seeks constantly to improve his instruction. Real mathematics teachers desire that their students, through proper and good instruction, become better prepared than in the past, to take their places in democratic society.

There are numerous avenues which a teacher can follow that will lead to improvement of instruction. The top-grade teacher will eventually have traveled all of these paths. One avenue is the study of society, its history, its development, its probable future status, and the role that mathematics plays in society. What is it that makes a society, and how does man function in this society each 24 hours of his life? How do you, I, and all others in the various careers react to one another? How do we carry on our physical, social, intellectual, and esthetic relationships? How can the school, and the mathematics instruction in particular, prepare youths to take their places in this society according to their several talents and interests? The study of these questions can provide tentative answers pertinent to the improvement of mathematical instruction.

Another avenue, not always too well traveled, is the study of mathematics—the knowledge of the field wherein we instruct. Not only is such knowledge being extended daily, but any new study of previously pursued knowledge results in new insight, new discoveries of facts and relationships, and with this a command of our field of instruction that creates a genuine feeling of confidence and authority in the knowledge we impart. Today teachers of mathematics must know more mathematics, and more about its application to aeronautics, transportation, navigation, finance, and economics than ever before. This means we must continue as scholars within our field from arithemetic to the most advanced structures of mathematical knowledge. This study cannot help but have an effect that will change the stress and the material we use in our teaching, and thus improve the attainment of goals more consistent with modern living.

There is the avenue of self-study, by reading, in modern thought. Too many mathematics teachers are narrow in their reading. They should come to know the modern thinking in philosophy, anthropology, physiology, geography, statistical methods, physics, and in general all those areas of knowledge that study human individuals as elements of larger populations. Four five, or even six years of college training can no longer be considered sufficient to provide complete preparation for teaching. So long as new knowledge is being developed in fields related to his profession, the teacher must be alert to its implication for his instruction in mathematics. A teacher, well-read in the other great areas of human knowledge, gains a perspective of the place of mathematics that puts it in proper focus in the whole educational program.

Many other avenues lead upward, in particular the fields of guidance and evaluation. What mathematics is a child at a given stage able to learn? How can this instruction be made available at the proper time? How much mathematics should a particular individual study, and what are the criteria that determine the type and the amount? At any given stage, how can we tell what achievement a student has actually attained in skills, in concepts, and in problem-solving? The study of proper guidance and evaluation can give insight into needed curriculum and teaching procedures that can produce better mathematical learning, better adapted students, and happier ones. Guidance and evaluation will contribute much to the improvement of instruction.

Another avenue is the study of human growth. Each succeeding year we learn more and more from biologists and psychologists of the physical and mental changes that take place as children grow into adolescents and into adults. The teacher seeks to know more of the causes of certain emotional behaviors, of the development of attitudes toward things and toward other people, both classmates and adults. How do certain likes and dislikes of the opposite sexes develop, and how will we secure good social behavior as the youngsters grow up? How can we increase the feeling of sympathetic personal relationship between student and teacher, a feeling so necessary in developing the personality of our youth? With increased knowledge of physical and psychological change in maturing youth, there is bound to be change in our procedures of instruction that will result in improved educational outcomes.

Closely paralleling the avenue of growth is the one with which this book has been concerned, the study of how we learn. In the analysis and rethinking of the preceding chapters, the teacher must continually bear in mind that the presentation is only one of the many avenues to better mathematics teaching; that is, an avenue cutting across all the other avenues of improved instruction which must be related to them at all times. Eventually the teacher must create a balance among learning theory, subject matter, the student, classroom practice, and educational goals. This balance can come only through study and experience, through knowing and applying—that is, through creating a behavior, a way of acting, that responds naturally and efficiently to mathematical instruction. This chapter is concerned with showing how learning theory can be integrated with the other aspects of instruction.

MEDIA OF COMMUNICATION IN LEARNING THEORY

The literature and addresses on mathematical education are replete with terms of psychological implications. We read of attitudes, appreciation, mental ability, memory, forgetting, association, thinking, and concepts. In many cases no attempt is made to define these words, and as a result there is vagueness and misunderstanding in the interpretation of the writing and the speaking. Thus learning can be conceived of vaguely as something that takes place in classrooms or in laboratories, or as acquired knowledge, or as something else. To avoid such ambiguity and to provide media for accurate communication in the theory of learning, it is necessary and sufficient that the terms used convey unique meanings in given contexts. This meaning should be common in the minds of all who hear or read the word to the extent that discussion or conclusions involving the concept can be carried on in a rational manner. Without such understanding, it would be impossible to put the psychological concepts to effective use in classroom instruction.

This book has provided the clarification of most of the terms used in psychological discussions on learning. By accepting the definitions, descriptions or illustrations herein presented, we have a medium for efficient communication of our ideas about learning. Other definitions could have been given, but those used in this Yearbook are either commonly accepted in education, or are the emerging concepts in modern thinking about learning. Thus when we speak of learning, we mean a definite change in behavior of the organism. Attitudes are not habits, but they are a developed set of the entire organism that makes it act or tend to act in a given way, when confronted with a given situation. Similarly, when such words as goal, skill, concept, motivation, problem-solving, drill, interest, and such arise in conversation, or in reading, they should convey the fairly exact meanings that have been established in these pages. For the purpose of studying learning it is as necessary to have these commonly accepted meanings of psychological terms as it is to have common concepts of geometrical terms in the discussion of properties of space.

This Yearbook has developed and clarified the meanings of the symbols and words used in learning theory. We now know what it is we are communicating, even though we do not know the complete nature of the concepts we use.

The interrelationships of the aspects of learning concern us. In most units of instruction the teacher is aware of the importance

THE IMPROVEMENT OF INSTRUCTION   339

of sensory experience, concepts, language, skills, problem-solving, motivations, individual differences, and transfer. He knows that these aspects of learning are interrelated and that there is no fixed sequence in which they are to be considered. Teachers differ greatly, however, in their points of view concerning the relative importance of emphasis to be assigned to each. First let us reexamine these aspects of learning.

What is the role of sensory experience? To one teacher (perhaps in the primary grades) it is to help the pupil in building concepts, in discovering facts or relationships; it is the initial step in proceeding from the concrete to the abstract. To another teacher (probably in solid geometry) it is to show, to illustrate, to make concrete that which the pupil already has learned. Obviously many sensory aids may serve the purposes of both types of teachers. In general, the teacher whose objective for using sensory experience in a particular learning situation is clearest will make the best selection.

In Chapter III we saw that effective learning requires the building of ever wider and broader concepts. The concept of a division as part-taking must be extended to include division as comparison. The once separate concepts of division and fractions must be merged. The concept of integral multipliers must be extended to include fraction multipliers. The teacher of algebra carefully extends the concept of exponent from positive integers to zero, to negative integers, to fractions. Progress through algebra may be described as an extension of the concepts of arithmetic. The extension of the concept of ratio from "total distance traveled divided by distance traveled in a unit of time" to $\frac{\Delta x}{\Delta y}$ dramatizes the meaning of broadening a concept. Clearly, mathematical learning is of necessity concerned with extension of concepts!

The interdependence of language (words, signs and symbols) and thinking was analyzed in Chapter VI. Through effective association of language and experience, one ultimately learns to work with ideas. What the pupil does and says and hears (and reads) in connection with finding the height of the school flagpole by indirect measurement determines his concept of "tangent

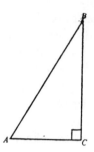

of an angle." He estimates its height, measures the length of its shadow and the angle of elevation of the sun, notes the right triangle, makes a scale drawing of the out-door-triangle, uses a notation for the triangle, measures the line in the drawing which corresponds to the height of the flagpole, compares, by division, BC with AC, finds that BC is about 1.7 times AC, and concludes that the flagpole is about 1.7 times as long as its shadow. Strategic in this total experience are the words and/or symbols: height of flagpole, length of shadow, angle of elevation, ratio of BC to AC, the side opposite the 60° angle divided by the side adjacent to it, 1.7. Then, after skillful guidance, the learner senses, understands, and says, "In any right triangle having a 60° angle, the ratio of the side opposite the 60° angle to the side next to it is approximately 1.7." Finally he is ready for the symbolic statement "tangent 60° = 1.7." Clearly, experience activities and properly associated words are essential to the building of new concepts.

The teacher chooses the vocabulary best suited to create the new concept. For another illustration consider the building of the part-taking concept of division. The teacher uses carefully selected words as "share equally," "dealing out," "divide equally among," "finding the size of each of a given number of equal groups," and other carefully selected words. He avoids using the vocabulary of the measurement concept: compare, measure, how many times, what part of, finding the number of groups of a given size, the divisor goes into the dividend, and other words of measurement concept. A concept emerges from a variety of carefully selected languages of a context.

But mathematical learning involves skills as well as concepts. The acquisition of skills in performing operations is greatly facili-

THE IMPROVEMENT OF INSTRUCTION

tated by understanding the concepts of the operations. The technique of "carrying" in addition becomes rational as a consequence of the associative principle. Obviously skill in mental addition is easily acquired by the order of addition shown in (a). The order of addition shown in (b) easily leads to an understanding of the technique of carrying shown in (c).

```
(a)   27            (b)   27            (c)   27
      45                  45                  45
      ──                  ──                  ──
      60                  12                  72
      12                  60
      ──                  ──
      72                  72
```

The pupil who considers addition as a process of putting together like groups and who thinks of a fraction as "one or more of the equal parts of a whole" will convert $\frac{2}{5} + \frac{4}{5}$ to $\frac{6}{5}$ rather than to $\frac{6}{10}$. Skills operating without the guidance of concepts often lead the learner astray.

Recently, inquiring teachers have been engaged in a "why-how" or "how-why" controversy. Which does come first, the concept or the skill? Or do they grow somewhat simultaneously? Available evidence indicates that skill learning should follow if not accompany concept learning. The older practice of emphasizing skill learning with little emphasis upon meaning led to the justifiable criticism that instruction was mechanical, that pupils learned the how without the why. Fortunately the influence of the field psychology, with its emphasis upon concepts and relationships, is increasing.

Students of mathematics education agree that power, ingenuity, resourcefulness in problem-solving is a major objective of instruction. They interpret problem-solving as finding by reasoning a satisfactory response to a situation which is novel, for which there is no available recalled response. In geometry a problem is presented when the pupil wonders whether two sides of a triangle are equal, if the angles opposite those sides are equal. Or,

"Can I show geometrically that $\sqrt{ab}$ can never be greater than $\frac{a+b}{2}$?" In algebra the pupil who wants to know whether A in the formula $A = bh$, is increased 25% when b and h are each increased 25%, is confronted with a *problem*. And in sixth-grade arithmetic, the pupil who would like to know how many apples he would have to buy at 3 for 5¢ and sell at 2 for 5¢ in order to make a dollar, has a problem. The adult who wonders what rate of interest is being paid when a refrigerator which can be bought for $200 cash is bought by paying $20 down and $20 monthly for 10 months is face to face with a problem.

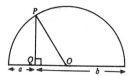

Now let us examine the solution of the second geometry problem stated above: Can $\sqrt{ab}$ ever be greater than $\frac{a+b}{2}$? The pupil who interprets $\sqrt{ab}$ as the geometric mean of a and b, may think of a right triangle inscribed in a semicircle (the situation in which he first met the concept of geometric mean). Thus the altitude PQ is equal to $\sqrt{ab}$. But what about the $\frac{a+b}{2}$? Oh yes, that is half the diameter; it is equal to the radius, PO. Obviously now $PQ \leq PO$, and hence $\sqrt{ab} \leq \frac{a+b}{2}$. The secret to the solution lies in seeing how, in what context, $\sqrt{ab}$ is related to $\frac{a+b}{2}$.

In general the secret of success in problem-solving is in seeing how what one wants to find is related to something in the problem situation that is known. Reasoning means discovering heretofore unrecognized relationships. The more relationships one knows, the less likely he is to have to resort to reason in successfully responding to a new situation. It must be apparent that recurring problem situations cease to be problems; the learner is able to recall the relationship required to solve the problem. Effective teaching transforms problem situations into familiar situations to which the learner responds without delay and without confusion.

Learning does not take place without a challenge. Doing the same thing over and over is not changing behavior. A student must continuously be faced with problems. This means the classroom situation must be carefully planned day by day so that new and more advanced aspects (but not too advanced) are being faced by the students. Even a review lesson should contain the concepts to be studied in new and varied siutations. The measurement of an angle by the arcs in which it intersects a circle can be studied in terms of positive and negative arcs as suggested in the figures below.

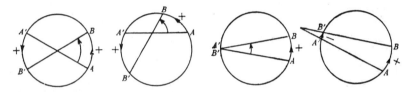

The relationship between positive and negative numbers, and positive and negative arcs as shown in the above figures is not apparent to all students. We cannot expect that all children will discover relationships entirely on their own initiative. For most students *learning needs direction*. If students are not making progress we give clues. To give the right kind of clues is skillful teaching. Perhaps a sentence "Let us agree that if $\widehat{AB}$ is positive then $\widehat{A'B'}$ will be positive only if it takes the same circular direction" will be sufficient for some students, but for others even stronger clues are necessary. Students should not be left entirely free to recapitulate the entire development of the human race; the teacher can guide them past unnecessary pitfalls. All that is essential is that the desired organized pattern of behavior (learning) must eventually be made by the learner for himself.

There is no one formula or procedure for problem-solving. Frequently the pupil discovers the strategic relationship needed to solve an algebraic verbal problem by assuming an answer, and checking it against the given conditions stated in the problem. Thus, in the problem: "A merchant bought 12 dresses, some at $8 each, and the remainder at $6 each. The 12 dresses cost $80. How many of each kind (price class) did he buy?" The pupil, seeing that the merchant must have bought less than 10 dresses at $8 each, may guess 5. Checking: 5 dresses at $8 each cost $40, the 12 − 5, or 7 dresses at $6 cost $42. But $40 + $42 is $2 too much. Hence, replace one of the $8 dresses with a $6 dress. Actually the learner assumed for the moment that he bought 5 of the $8 dresses and determined, by using the conditions of the problem, that the guessed answer must be 5 − 1 or 4 of the $8 dresses.

Even though the solution of a problem often appears to come quickly as a result of sudden insight, it is more likely the result of analysis of relationships than the product of trial and error.

There are many correct, if not equally good, methods of problem analysis. In the apple problem, previously stated, one solution involved finding the profit on one apple ($5/2¢ - 5/3¢ = 5/6¢$) and then dividing the desired profit, 100¢, by the profit on one apple. Another solution, more mature and ingenious in conception, first finds the profit on a group of apples (6 or 12 or any small multiple of 2 and 3). If a group of 12 be chosen for study, the profit would be 10 cents; for a profit of 100 cents, 10 such groups would have to be bought and sold.

A less mature, more stereotyped solution is likely to be proposed by pupils in the algebra class: Let x represent the number bought; then $5/3 x$ is the cost and so on.

Pupils achieve greater power in problem-solving by proposing and evaluating alternate methods of solution of a limited number of problems, than by solving a greater number of problems by any one pattern of analysis. Teachers of geometric "originals" have long been convinced of this belief.

Our students should not only be able to reason and to solve problems, they should come to enjoy them. No teacher is unaware of the potency of interest or motivation in promoting learning. Pupils learn best when they want to learn, feel a need for learning, when they are curious to find out.

Psychologists are skeptical concerning the existence of special aptitudes for mathematics rather than another school discipline, but they are certain that for motivation nothing succeeds like success in thinking. The pupil who fails three or four times out of five tries soon loses interest; the pupil who never fails soon finds no challenge and loses interest; but the pupil who succeeds three or four times out of five tries likes the adventure and wants more of it.

The teacher who likes mathematics, who is able to arrange learning exercises of such difficulty that the pupil experiences success more often than failure in his thinking, and who exhibits genuine pleasure when the pupil succeeds, need have little concern about the problem of motivation. Successful intellectual behavior generates motivation.

A group of pupils, as well as the individual pupil, thrive on success. Group thinking (currently known as group process), with little apparent direction of the teacher, creates interest and gives to each member of the group a sense of belonging and achievement. The traditional recitation (question by teacher, response by a particular pupil) is replaced by discussion, planning, appraising, by procedures in which the individual pupil contributes according to his interests and talents.

The two preceding paragraphs contain far-reaching implications for handling the problem of individual differences. Few teachers today would wish to replace group instruction by tutorial instruction. They prefer, instead, to replace random (alphabetic, often) selection of groups having a wide range of ability and achievement, with more homogenous groupings. Instructional materials as yet are inadequate for meeting the problem at any grade level.

Teachers know that very few of the problems with which a learner may ultimately be confronted can be considered in school. Thus they constantly turn to the psychology of transfer for guidance in making the learning as general as possible. The psychology of "transfer of training" is replete with implications for the teacher of mathematics. In order of their extent of transfer, skill learnings rank lowest. Concepts are susceptible of significant transfer. Most widely transferable however are the attitudes, likes or dislikes, confidence or insecurity, the emotional concomitants.

And the least enduring learnings, those eroded most by disuse and forgetfulness, are the skills, the facts, the techniques of mathematics. Concepts are far more enduring. Years after a successful experience with deductive proof in the secondary school the adult retains his concept of proof. But most permanent are those aspects of learning which we are wont to call appreciation, pleasure, frustration, attitude. How a pupil feels about his course in geometry for example may significantly affect his adult behavior as a parent, or as a member of the board of education.

SOME PRACTICAL CONSIDERATIONS

The arrangement of the chapters in this book was not without purpose. There was first presented an orientation to the general theory of learning—experience within the field being studied. Then some of the main elements of the field-motivation, attitudes, concepts, and sensory impressions were treated so as to give deeper insight into learning. The relationship of these elements through language, practice, and transfer, paved the way for a re-examination of learning as problem-solving. In order to prevent any misinterpretation of learning as unguided experience of novices, the particular precaution was taken to show the need for providing for individual differences and for growth through planned instruction. This final chapter has been concerned with creating a total picture of these aspects of learning—their inter-relatedness—to enable the reader to achieve a rather complete and balanced configuration that can aid in adapting learning theory to good classroom practice. To avoid some possible gaps or flaws, a few further practical considerations seem pertinent.

Children do not grow and develop in mathematical knowledge in a vacuum of such knowledge. Quantitative thinking is acquired in active mental dealing with quantitative situations. The quantitative aspects of a situation can go unnoticed unless they are deliberately brought to the sphere of attention of the student. In the study of health, cost of medical care, cost or nature of medical insurance, span of life in various occupations and sections of the country, the relation of climate conditions (quantitatively measured) to types of diseases, and so on, are the means of insight to the need for mathematical learning. The necessary mathematical concepts and skills in treating such problems must be developed by actual experience in organized, planned mathematics instruction. The child grows and develops in a numerical and geometric environment in which he changes his behavior to ever more complex organized patterns.

But progress is not made along a straight line from the fundamentals to the generalized theorems. This is the way the final learning can be organized, and is frequently so organized in texts, but it is not the way most students grow in wisdom. Our initial learnings in algebra are not Peano's axioms, nor is our initial learning in geometry Hilbert's postulates. Rather the initial learning is in an area in which exploration first takes place. We count, measure, draw, and make preliminary statements which are refined downward toward the foundations and upward toward more abstract, complex, and generalized relations. After a while we discard many particular theorems for one more generalized theorem, and in advanced stages of learning only is the area of exploration reorganized as a straight line mathematical development.

The growth in mathematical knowledge by each individual student thus calls for direction by a skillful teacher who has a balanced emphasis on the various phases of learning. He uses the heuristic method only so far as pupils need directed questions in the quest of their learning. He has an experimental attitude that allows freedom of approach in learning to the degree that the experiment is headed toward a desired concept, or relationship, and directs the experiment back to fruitful approaches when the student is adrift. He recognizes that drill can be dangerous and boring as well as good habit formation procedure. He balances each step from concrete experience to semiconcrete representations, to words and symbols, to generalized abstract theorems through proper evaluation and by appropriate attainable challenges. He also recognizes that not all learning begins in concrete material objects, but that much new learning may start in already learned abstractions. Thus algebraic fractions are referred back to the abstractions learned in the study of arithmetic fractions and not back to parts of concrete objects. Geometry is related to geometrical drawings, trigonometry is related back to geometry and algebra.

It is just because of this sequential aspect of mathematical knowledge that proper balance becomes the all important aspect in its learning. Unless a clearly and correctly formed mathematical concept emerges from a learning situation and it is related to other phases of already learned mathematics through sufficient practice to gain skill in the use of the concept, the learning of all later mathematics dependent upon this concept is seriously (and sometimes totally) impaired. The lack of correct concepts in arithmetic may be one of the great reasons for the difficulty algebra presents to so many of our students. It is this balance that is indirectly referred to in every chapter in this text: a mathematical problem within the comprehension of the student, but still a problem as the start of learning; motivation, sufficient to send him on toward the solution; the challenge to abstract, to generalize, to form concepts; the proper use of sensory aids, of drill, of appropriate language; not doing too much nor failing to do enough; and adapting these measures to the various individual differences in learning. To gain such balance is to become a master teacher—the goal of all professionally-minded educators.

It should be evident by now, that the position taken by this book is: We learn that mathematics which we are taught. The teacher is the primary agent and force in the learning that takes place. The teacher is all important. There are some persons who say one who knows cannot teach for he cannot fathom the difficulties of his students. These persons say that as a teacher works with his students through a problematic situation which is new to both teacher and students, real learning takes place and then only. We believe this assumption to be entirely erroneous and assert that a teacher is a learning engineer, a builder of minds that will solve problems. As such, the teacher must first know the total mathematics he will teach, that is the framework, the facing material, and all the interior finishings to make the structure complete. But he must also know the stresses and strains the foundation and framework must support, and how the whole structure of mathematics is put together in the minds of his students. This story of mental construction has been put forth in this book as a practical guide to the classroom teacher, the learning engineer. It is the teacher who is responsible for planning, designing, blueprinting, constructing, finishing, and putting the final product—a problem-solving mind in the realm of quantitative thinking—to effective use in society.

It was shown at the opening of this chapter that there are many avenues to improvement of instruction. How we learn is only a part of the whole organization of teaching. We must weigh its importance in its relation to the other avenues of learning. Also, the present theory and practice of learning is not a fixed body of knowledge of the type one finds in certain mathematical disciplines such as plane Euclidean geometry. Until we know more, we must conceive of the aspects of learning—motivation, analysis, transfer of training, and practice—as fluid elements, as tenative workable explanations of a theory of learning. When we gain more knowledge of the operation of the brain, and of the manner in which human behavior is changed, these various aspects may shift position, change in their importance, and even new elements may enter the picture. The alert teacher will be on the outlook for new developments and new interpretations and balance them with his present conception of how learning takes place, a concept that this book has given.

Thus in a sense the teacher must be a master technician. He must know how to build any known kind of learning. But he also must be a philosopher. He must decide what kind of learning is of most worth. He must weigh, balance, and appraise the possible learnings. He must know their relative worth both for the individual and for society. He is at the moment following the recent discoveries in psychology, in the process of readjusting his thinking about the relative emphasis to be assigned to skills, concepts, problem-solving, and attitudes. We predict that instruction in mathematics for general education will come to rely more and more upon the implications of the following:

Learning is thinking.

Successful thinking is possible at any grade or achievement level.

Successful thinking is heavily dependent upon concepts and relationships.

The satisfactions following successful thinking provide enduring enrichment for the learner.

INDEX

Abstracting, 172
Abstraction
  in concept formation, 82–83
Accuracy, 57–58
Action system of learning, 88
Aesthetic appreciation, 110
Affiliation, 59
  need for, 46
Aim of lesson, 307
Analogy, 259
Analysis
  in problem-solving, 259
  in thought, 33
  law of, 14
Appreciation, 110
Approval, 59
  need for, 46
Aptitude in mathematics, 345
Assignment
  planning of, 323 ff.
  spiral type, 323–25
Attention
  improvement of, 119
    by controlling observer, 138
    by controlling stimulus, 137–38
Attitudes, 45
  development of, 111
  in achievement, 55–56
Average children, 275

Balance
  in instruction, 337
  in teaching, 347–48
Basic lesson planning, 307
Behavior
  as learning, 4
  change in, 2–3
  experimental study, 2
  physical, of brain, 1
Belonging sequence, 14
Board work
  planning of, 320
Bond, 13
Bright children, 275

Challenge in learning, 343
Checking, 57

Color in concept formation, 78
Communication media, 337 ff.
Competition
  as motivation, 62
  effects of, 62
Concepts, 9
  application of, 107
  attainment of, 82 ff.
  formation of, 69 ff.
  nature of, 77
Concept building, 339
  example of, 106
Concept formation, 69
  examples in, 78, 79
  implication of, 92
Conditioning, 11 ff.
  characteristics of, 12–13
  fundamental principle, 12
Configuration, 22
  aspects of, 23–24
Connectionism, 13 ff
  characteristics of, 17
  laws of, 13, 14
Context, 170
Curriculum and attitudes, 58

Deduction, 35
Development of lesson, 315
Devices
  interest arousing, 51–53
  limitations of, 52
Diagnosis, 265
Diagnostic testing, 275
Diagrams in problem solving, 252–53
Differentiation, 321
  in courses, 296
  in depth and scope, 285–86
Discovery, 260
Dramatization, 254
Drill
  definition of, 192
  examples of, 193–94, 198
  in algebraic equations, 201
  in fractions, 199
  in lesson planning, 321
  principles of, 195–96
  uses of, 197–98

Effect
 law of, 13
Einstellung, 243-44
Emotional activities, 110
Emotional learning, 148
Enrichment, 277
Environment, 263-64
Equations, 202
Estimation
 improvement of, 120
 use of, 120
Exercise
 law of, 14
Exercises, 243-45
Experience
 in concept formation, 78-79, 94-95
 in problem-solving, 250, 260
 reorganization of, 215
 successful, 56

Failure, 56
Field theories, 18 ff.
Formal discipline, 205-207
Formula
 application of, 262
Fractions
 computational skills with, 200
 concept of, 199
 functional competence with, 201
 principles of, 199

General mathematics, 297 ff.
Generalization, 21
 doctrine of, 212
 in concept formation, 81-84
Goals, 43-44, 47
 life, 47-48
 near, 312
 realism in selecting, 50
 substitute, 312
Group activity, 63
Grouping, 274
 by classes, 291
 illustration of, 292-93
 in classes, 279-81, 290
 in geometry, 294-95
 limitations of, 291
Growth of children, 337
Guidance, 336-37

Heuristic method
 in lesson planning, 317 ff.
 in problem-solving, 256
Hypotheses
 forming, 244-46
 framing, 34
 testing, 247, 267-68
 value of, 267

Identical elements, 207 ff.
Illumination stage, 39
Imagery, 112
 improvement of, 149
Images
 types of, 112-14
Imagination, 114, 115
Incentives, 44
 to learning, 59, 60
Incubation period, 38-39, 259
Individual
 active role of, 84-85
 passive role of, 84
Individual differences, 271 ff.
 adjusting to, 273
 history of, 271
 in grouping, 345
 in problem-solving, 251
 table of, 274
 traits, 271
 variation in grade, 272
Induction, 260
Inhibition
 associative, 12
Insight, 19
Instruction
 improvement of, 335
Integration, 322
Intelligence
 as adaptation, 4
 as understanding, 4
 distribution of, 275-77
 mechanical, 5
 nature of, 4 ff.
 social, 5
Interests, 45, 51

Language
 examples of strength, 157
 examples of weakness, 156-57
 in concept building, 340

INDEX

in mathematics, 156
in thinking, 339
mathematical, 158
Language symbols, 92
Law
of analysis, 14
of effect, 13
of exercise, 14
of readiness, 14
Laws of psychology, 312
Learning
animal, 10
areas of agreement, 29
as analysis, 19
connectionism, 16
creative, 37–39
definition of, 99
diagram of, 7, 37
directing of, 343
end-products, 9–10
enduring, 346
explanation of, 32–37
general principles of, 30–32
gestalt, 20–22
levels of, 278
problems of, 40
study of, 1, 2, 337
theory of, 1 ff.
Learning aids, 54
Learning mathematics, 24 ff.
Learning situation, 6–7
Lesson
basic, 306
drill, 332
example of, 327–30
types, 318
Lesson development
logical, 319–20
psychological, 319
Lesson planning, 331 ff.
Lesson plans, 303–304
complete, 326–27
Lessons
groups of, 333
special purpose, 331
Levels of learning, 279–80, 285
Library
use of, 222
Linear equations, 202

Maps, 178
Marks
as evaluation, 61
as incentives, 60–61
Mathematical knowledge, 347
Mathematics
as thinking, 158–59
study of, 336
Maturation, 212
Meanings
by agreement, 164
example of, 76
in motivation, 63–64
meaning of, 70, 75–76
of a problem, 248
of mathematical terms, 166
of symbols, 160
Memorizing, 104–105
improvement of, 144
Mental association, 9
Mental concepts, 106
Modern thought, 336
Motivating
criteria for evaluating, 65–66
Motivation,
definition of, 104
examples in lesson planning, 313–15
improvement of, 143
in mathematics education, 42 ff.
planning of, 310–11
Motives, 43
acquired, 43
inborn, 43
useful in teaching, 44
Motor skills
development of, 117
examples of, 117 ff.
Multiple solutions, 263
Multiplicity of meaning, 164
Multi-sensory aids, 86–87

Natural closure, 312
Need, 7
Non-academic courses, 301
Non-verbal signals, 160
Number concept, 87–88
Number ideas, 91

Objectives
  of lesson, 307–308
Order, 58
  of concept formation, 85–86
Orientation to problems, 252
Organization of relationships, 35

Perception, 102
  improvement of, 140
  of movement, 142–43
  patterns of, 102–103
Perceptual discrimination, 120
Perceptual illusion, 125
Perceptual-motor skill, 9
Planned instruction, 303 ff.
Planning
  need of, 303
  relation to psychology, 306
Planning instruction, 304–305
Practice, 192
Pragmatic dimension, 70
Prediction, 267
Pre-equation learning, 201
Preparation stage, 38
Problem
  analysis of, 344
  clarification of, 239
  definition of, 228
  individual meaning of, 238–39
  meaning of, 230–31
  orientation to, 237 ff.
  solving of, 229
Problems
  adaptation to levels, 283–84
  function of, 234–35
  number of, 233
Problem situation, 7, 32, 37
Problem-solving, 9, 228 ff.
  analyses of, 236
  examples of, 342
  framework of, 237
  importance of, 233
  improvement of, 147, 247 ff.
  steps in, 229–30
Productivity of thought materials, 255 ff.
Psychology in drill procedures, 194 ff.
Punishments, 60
Purposes, 44–47

Quantitative thinking, 346
Questioning, 316–17
  attitude, 254
  in problem-solving, 257

Reaction, 86
Readiness, 201
  law of, 14
Readings in problem-solving, 250
Reasoning
  development of, 89
  in mathematics, 159
Receptors, 99–100
Recreations, 53
Recurring experience, 192
Referents, 71–72, 174
Relationship, 260
Remedial instruction, 275
  in problem-solving, 265
Rewards, 60
Root, 162–64

Science, 171
Search model, 239–40
Security, 57–58
Self-esteem, 59
  need for, 46
Semantic dimension
  examples of, 72–73
  meaning of, 71
Semantics, 166 ff.
  principles of, 170 ff.
Sensations, 100–101
Senses
  definition of, 99
  improving use of, 116 ff.
  types of, 99
Sensory experience, 339
  in concept formation, 80
Sensory learning
  applied to mathematics, 99 ff.
  in developing concepts, 106
  in memorizing, 105
  in motivation, 104
  in problem-solving, 107–109
  psychology of, 102–103
  relation to other aspects, 103 ff.
Sensory-motor skill, 9

INDEX

Set
  in problem-solving, 242, 258
Skill
  acquisition of, 341
  versus concept, 341
Slow children, 275
Society
  study of, 335
Spacial perceptions, 141
Span of apprehension, 241
Step procedures, 265
Stimuli, 101–102
Success
  in motivation, 56–57
  in problem-solving, 264
Summary of lesson, 308
Superstitions in mathematics, 111, 112
Supplementary aids, 310
Symbolic response, 81
Syntatic dimension
  examples of, 72–73
  meaning of, 70
System, 58

Teacher
  as psychologist, 249
  as technician, 349
  importance of, 348
Teaching aids, 65
Teaching by wholes, 286–87
Teaching for meaning, 187
Teaching of mathematics
  cutaneous sensations, 132–33
  gustatory sensations, 132
  kinesthetic sensations, 132–33
  muscular skills, 135–36
  olfactory sensations, 132
  organic sensations, 132–33
  sensory learning, 129 ff.

Temporal perception, 142
Terms
  psychological, 338
Testing hypotheses, 266 ff.
Thinkers, 167–68
Thought
  complete act, 32 ff.
  reflective, 236
Thought material, 240 ff.
Transfer
  examples in teaching for, 222
  factors of, 205
  maximum, 217 ff.
  order of, 345
Transfer of training, 205 ff.
Trial and error, 15

Understanding, 180
  example of, 75
  in motivation, 63–64
  meaning of, 74–75
Understanding of concepts, 145

Verbal instruction
  in concept formation, 93
Verbalization, 183
  in problem-solving, 252
Verification, 36, 267
Visual aids
  as motivation, 48, 54
  use of, 93–94

Way of life, 170
Whole situation, 19
Words
  arbitrary nature of, 165
  as symbols, 160